Rationality and Science

B

Rationality and Science

Can Science Explain Everything?

Roger Trigg

BLACKWELL
Oxford UK & Cambridge USA

First published 1993

Blackwell Publishers
108 Cowley Road
Oxford OX4 1JF
UK

238 Main Street
Cambridge, Massachusetts 02142
USA

British Library Cataloguing in Publication Data

A CIP catalogue record for this book is available from the British Library.

Library of Congress Cataloging-in-Publication Data

Trigg, Roger.
Rationality and science : can science explain everything? / Roger
Trigg.
p. cm.
Includes bibliographical references and index.
ISBN 0–631–19036–8. — ISBN 0–631–19037–6 (pbk.)
1. Science—Philosophy. I. Title.
Q175. T74 1993
501—dc20   93–17136
                                                      CIP

Typeset in 11 on 13 pt Baskerville
by Graphicraft Typesetters Ltd., Hong Kong
Printed in Great Britain

This book is printed on acid-free paper

Contents

Acknowledgements

This book was largely written in the course of study leave granted to me by the University of Warwick. I spent a year as a Visiting Fellow at St Cross College, Oxford, and wish to record my gratitude to the Master and Fellows of St Cross for their kind hospitality. I much value my association with the College. I am also grateful to the British Academy for the personal research grant I was awarded to help me with some of my costs.

I have discussed the contents of this book with too many people to acknowledge individually. I am, however, especially grateful to Charles Taliaferro, whose sabbatical leave in Oxford coincided with mine. His advice and comments have been invaluable, and I am sure that the book has been greatly improved as a result.

Introduction

The Public Arena

In the Middle Ages in Europe the Christian Church was the
guardian of truth, or at least of what was accepted as truth. Yet
the doctrines of the Church gained their apparent potency not
simply because of the political power of the institutions proclaim-
ing them. Rather, their authority was believed to have been
grounded in the very nature of things. Ultimately everything was
alleged to be traced back to God the Creator. The Pope's authority was a devolved one, since as Vicar of Christ he could claim to
be representing God's own revelation to humanity. The Church's
hold on truth was alleged to stem from its insights into a reality
that was totally independent. Cynics would claim that an appeal
to objective reality was no more than a subtle move in its lust for
power. With God on one's side, one hardly needs any further
legitimation. Yet, as we shall see in this book, it still remains the
case that any claim to knowledge must be grounded in the way
things are.

The Reformation and the growth of modern science together
began to undermine the Church's appeal to authority. Even
Christians became suspicious of the abuse of power in the Church,
seen as a human institution, and it fragmented. Gradually it
was recognized that there were other ways of establishing truth.
Through the seventeenth and eighteenth centuries human

reason came to be seen as capable of discovering the truth about the world we live in, and the methods of the new science were thought to be particularly appropriate. Above all they seemed capable of producing agreement though accepted procedures. It was hard, for instance, to view a telescope as an instrument of social control by anyone eager for power. It merely enabled people to see what they previously could not. Furthermore it was possible for anyone to look as long as they had access to the instrument. Knowledge could not be restricted to a privileged few. Anyone could see the mountains of the moon as Galileo did. Inevitably many began to distrust large metaphysical schemes, devised by a few, to tell the rest of us what had to exist. Instead they preferred to start with what could be shown to be knowledge through piece-meal investigation of the world around us. Experience gradually became the touch-stone of truth, but it was not private experience. It could be shared. Science was well and truly in the public sphere, dealing with apparently objective matters which could be rationally established. The scientific and the objective became so associated as to seem virtually synonymous. What was objective could be publicly shared. A corollary was that scientists dealt with 'facts' while everything else was consigned to the personal, indeed the subjective, realm.

The empiricist, David Hume, was able in the eighteenth century to make the distinction between what is the case and what ought to be the case. He went on to say that 'morality is more properly felt than judged of'. Questions about virtue and vice were thus consigned to the private, subjective world. It was no accident that Hume was also unable to make any room for the notion of the self. The paradox was that with the emphasis on the public role of reason, soon to be modified in modern scientific method, anything private or personal would inevitably become problematic and dismissed as subjective, at best, and non-existent at worst. The more that everything of importance was thought to be objective, the less grip was left on the idea of individuals exercising their reason. With the modern conception of objectivity goes the view that what is objective must, in principle, be agreed. The idea is that people in similar situations will experience the same things and should have no difficulty in reaching agreement. What is contentious and cannot be settled by publicly acknowledged procedures is hence thought to be subjective.

The public arena is thought to be the preserve of science where truth can be objectively established. Reason and scientific method become progressively identified, and all else is swept into the category of the subjective. 'Values' become of questionable status. When the world was seen as God's creation, the idea of an ultimate split between facts and values would be unthinkable. In God, what is and what ought to be were seen to meet. Yet if judgements about what ought to be the case are merely personal preferences, they are clearly beyond the scope of science. Morality, aesthetics and religion are thus matters of subjective concern, because they could make no claim to scientific validity nor command general agreement. Indeed it was ironic that the more that the subjective world was populated with judgements that had previously been regarded as of immense importance, the less the concept of the subject was emphasized. By definition science cannot cope with the subject, since it is not accessible to the public examination demanded by the constraints of scientific objectivity. The impersonal, detached view of the scientist might provide a model for truth, but it left no room at all for any understanding of the individual consciousness that made it possible in the first place. Rationality came to be identified with the path to public agreement. It followed that when there was apparently irresolvable disagreement, reason was obviously not part of the picture.

The subjective, therefore, was not just unscientific but beyond the scope of reason, and perhaps positively irrational. Science presided over the public arena where truth could be rationally established, if not by general agreement, at least by experts who were generally accepted as much. The irony of this was that it could so easily lead to the development of a new priestly caste. Science with all its modern complexity, was in the end going to be no more democratic than the Church it had seemed to supplant. Individual judgements and preferences which varied between people were, at a theoretical level, excluded from its concern. Nevertheless they still had to be taken account of in

society, and so such differing values were tolerated in the demo-
cracies which were growing up as science advanced. At first in
England, it was just a question of tolerating 'dissenters' from the
Established Church, but as time went on a more radical pluralism set in. The United States, for example, was wary about allowing religion in the public sphere at all. As long as science remained
neutral about 'values', it could rise above such social tensions.
Indeed many thought that science itself was the key to real
progress, and that anything else was a distraction.

Facts and values, however, are not so easily sundered. Nothing
could be more 'value-laden' than the idea of progress, and yet
opposition to such progress seems to many not just intelligible
but positively desirable. Science has been pushed off its pedestal.
Far from being value-neutral, the whole-hearted pursuit of
knowledge in physical science, and its unthinking application,
has been the dominant value of recent generations. There is now
no guaranteed public acceptance of the role of empirical science. It no longer commands universal respect, and inevitably its
monopoly claims to knowledge are being questioned. It is now
blamed for everything from arms races to the pollution of the
planet, and the erosion of the ozone layer. The question of justification has to be faced. It is not enough simply to practise
science.

Once this is acknowledged, it is not enough to rely on the bare
fact of acceptance within the public sphere. Appeals to the 'objectivity' of science merely beg the question. Too much has been
consigned to the realm of the private and the subjective. For
example, the prevalent view that morality is a private matter must
be connected with the implicit connection between the public,
the scientific and what can be agreed. Once doubts arise even in
the public realm about the worth of science, this whole way of
thinking must be questioned. Science cannot simply rely on the
contingent fact that it has acquired and enjoyed social prestige.
It cannot be self-contained, but like the medieval Church, it must
rest its claims to authority in something beyond itself. It is the
argument of this book that science has itself to appeal to a
metaphysical basis. So far from science being the enemy of
metaphysics, it stands in urgent need of its assistance. Much, of

course, depends on what is meant by metaphysics. Science cannot, however simply be accepted on its own terms. It is not enough to be told that the world of science *is* our world, that scientific rationality *is* rationality. We need to be told why this is so. Above all, when people grow cynical of the world of science, we need to be able to give reasons as to why science should not be ignored. Too many philosophers and scientists are willing to take science at face value. It is so obvious to them that physicial science sets the standards for truth, that they do not allow the question of justification even to be raised. That is, however, an insecure position to be in, and it is too dependent on social factors. Once the assumptions of the modern world cease to be generally shared, science will find it stands in need of rational justification. That this is in fact necessary is the theme of this book.

Reason and Anti-Realism

The story will begin, as it must, in Vienna between the wars. It was then that the philosophy which was to sweep over the English-speaking world was fashioned in the so-called Vienna Circle. The methods and discoveries of modern science dominated it, and philosophy could be regarded as the mere handmaiden of the sciences. Scientific success and progress could be taken for granted, and 'science' was taken to be virtually synonymous with 'knowledge'. The 'scientific world-conception' dictated what was to be philosophically acceptable. Yet just because philosophy was relegated to a secondary role, the task of justifying the practice of science was no longer thought essential. Science defined what is meant by truth, and there was no room for wondering whether it was the only path to truth, or a path to it at all. Metaphysics became a term of abuse.

This wholesale identification of human reasoning with the methods of science was bound to be questioned. One reaction is to recognize the importance of science but to insist that rationality should not simply be identified with it. It could be argued that science itself stands in need of a rational underpinning.

That is the programme of this book. The other reaction possible is much more radical, since it questions the very notion of rationality as a path to truth. Instead of making science the measure of truth, it will say that there can be no such overarching standard. Human activity is divided into many different practices and each develops its own standards and way of dealing with matters. The practice of physical science will then be seen as one such way of life. Its criteria for what is to count as a reason are only internal, and it will be regarded as profoundly mistaken to apply those relevant to science to other ways of thinking and acting. Neither ethics nor religion would then be despised merely because they are not science. They, too, it will be said, have criteria of reasonableness. A moral judgement is not a scientific one, and neither are the same as religious claims, but none, it seems, will be better or worse for that. They are merely different.

The fragmentation of reason and its division into different compartments is a powerful way of resisting the scientifistic claim that science is all that matters. It is less impressive as a way of providing any form of justification. Reason always appears as internal to a practice or way of life. It can then never be abstracted from a particular context and used in more free-floating manner to justify or condemn whole patterns of life. We have instead to take the appropriateness of an activity for granted without examining its presuppositions. This mode of argument appears regularly in contemporary philosophy. It is particularly associated with the later views of Wittgenstein, who himself explicitly reacted against the views of the Vienna Circle. It is also readily associated with various strands of American pragmatism, which believe that we should always begin with actual practices, rather than the abstractions of metaphysics.

An underlying battle rages between realists who believe that there is a world to be investigated which exists independently of human belief and language, and anti-realists. The latter wish to build our conceptions of truth and reality on the way human language is integrated with a particular way of acting. Truth is constructed and not discovered, they would maintain. We start, not surprisingly, where we are and, build up our understanding of the nature of things from what we already take for granted.

There is always, therefore, a hidden reference to human understanding and human capabilities in the anti-realist approach. This argument always turns on the difficulty of having a conception of something beyond our conceptions. Reality is still something being conceptualized by us even when it is understood as being beyond our reach. Again and again we shall return in this book to the implications of various versions of this argument. Anti-realism appears in many different forms, even though often the same issues are at stake. In the same way, the problem of reflexivity will constantly appear. Anti-realist claims can never be finally grounded in reality. By definition, they must always stop with how we can envisage reality. This lack of grounding, however, can produce problems when global claims about human reasoning are made. To take a simple example, if all reasoning only takes place within particular social practices, what is the status of any reasoning *about* social practices as such? Is it merely the product of another social practice? If it is, it must be of little interest, since whatever force it may seem to have comes from its being understood to be making a claim about what is the case. The fragmentation of reason alluded to would mean that wide claims about reason, even those about its fragmentation, cannot be made.

This kind of example recurs as a regular theme throughout the book. There is a pattern of argument prevalent which seems to relish giving reasons as to why reason is impossible. The contradiction should be obvious, but the persistent temptation to fall into it demonstrates how the very possibility of rationality is held by many to be in question. In part this is because they also reject the idea of an objective reality which is the same for everybody, whether they recognize it or not. The physical sciences provide a test case for the application of rationality, but it is not just a question of being suspicious of the claims of science. Some deride all reasoning on the grounds that it is the product of particular historical periods, and intelligible only in its own context. This attack on reason is so pervasive in our culture that it can masquerade as a constituent part of our contemporary understanding of ourselves. 'Post-modernism', we are told, has rejected 'modern' conceptions of reason.

A paradox is that similar conclusions about reason can be

reached even from within science. A pre-occupation with scientific method can result in rejecting any notion of rationality which cannot be fully translated into physical terms. The nature of consciousness becomes problematic, and thus the idea of a rationality that cannot be reduced to the description of physical processes is made highly suspect. We apparently arrive at scientific reasons for rejecting the possibility of reason as it has been traditionally understood. This kind of fully-fledged naturalism or physicalism restricts reality to what can be fully described by the natural sciences. Yet in so doing, it too makes the practice of science ungrounded in any rational way. The problem of reasoning about rationality will be constantly referred to in this book. An inability to reason about the nature of reality in a way that can transcend the immediate constraints of history, society and our physical make-up undermines the very possibility of physical science. It is always possible to define rationality so that it implies no such abstraction from a particular context. The question, though, of the warrant of such a definition will immediately arise, and we are then plunged straight back into questions of rational acceptability.

Human nature must be related to reality. If the gap between the two is made too wide, scepticism soon arises about the possibility of any knowledge. If, on the other hand, the gap is closed so that we can be sure of knowledge, the effect is bound to make reality anthropocentric. Various philosophical and scientific theories have arisen to try to bridge what seems an uncomfortably wide chasm. Evolutionary epistemology has tried to give an explanation for the way our beliefs and theories appear to fit the world. In the same way, in cosmology various versions of the so-called 'anthropic principle' have been canvassed to show a link between our existence as humans and the character of the universe. There are different methods of demonstrating a link between us and our scientific beliefs on the one hand, and reality on the other. The urge for a complete scientific explanation of everything remains strong. At best it can be a stimulus to further scientific investigation. At worst, however, it can be merely a contemporary example of the urge to make science the final arbiter of what is real. Yet without some explanation as to why the human mind should be capable of such a feat, this must remain somewhat unconvincing. For science to explain everything, we need a reason for trusting science. This once again brings us back to the question of a rational grounding in reality.

The Character of Reality

Religion has undoubtedly played its part historically in providing an intellectual climate in which modern science could flourish. This may seem surprising to those who have been brought up to assume that science and religion were rivals. Nevertheless, a belief in a God who is the source of reason and the ultimate explanation for an inherent rational structure in the world does provide an instructive alternative model to that of a science which has to be accepted at face value. Certainly the apparent regularity and order of the world is thereby explained, though clearly it is not the kind of explanation that is acceptable to everyone. Resistance to such metaphysics was a motive for much of the avowedly atheistic work of the Vienna Circle. It is not part of the task of this book to argue for theism. The point is rather that there are some possible ways of validating science.

There is not an automatic choice between an ungrounded trust in science as the only source of truth and a relativist acceptance of science as one of many alternative forms of life. Other views can and do exist which give science a metaphysical basis. They seek to ground science without letting it have monopoly rights over our life. It is in this connection that belief in God is relevant. Whether true or false, it provides an interesting example of one of several ways in which science can be seen as a fairly trustworthy product of a human rationality, which is set in a reality that is not of its own making.

The quest for a justification of physical science is more pressing than it once was because it is no longer obvious that the development of science is to be equated with human progress. The benefits of scientific discovery are often overshadowed by the undoubted costs. Environmentalists become ever more hostile to what they see as a source of danger not only to humans but to the whole planet on which we live. They are not going to be convinced by a demand that we trust science, or that science is to set the standard for what is to count as knowledge. The risk is that they resist science by acquiescing in the notion that it is just one interpretation of the world amongst many. It is merely yet another form of life. This may cut science down to size, but it does nothing to enhance the claims of environmentalism. They too can easily be dismissed by others as members of some eccentric way of life. They must, instead, appeal to a concept of a nature which exists in itself, independently of human conceptions. In other words, environmentalists can only claim to be heard and only present a coherent theory by being realists. Anti-realists are in no position to talk of any nature that is radically independent of us. Their vision will always in the end relate all reality logically to human understanding. In fact, the very reverse approach is necessary. Physical reality is so far from being constituted by our concepts, that in fact we do not properly realize just what we are doing to the whole ecosystem of which we are part.

Human actions produce many direct and indirect consequences in the physical world, which arise largely as a result of scientific and technological advance. We may not choose to notice some of them, but others are difficult to ascertain. Those who are concerned for what humans are doing to the global environment have to take seriously the fact that what we think we are doing and what we are doing may be very different. Nature may well have the last word, but, if so, it is a nature that is totally independent of human concepts. It is not an unstructured chaos waiting for the human mind arbitrarily to arrange it. It has characteristics, even a life, of its own. Human interference may disrupt it and even destroy it, but it is certainly not constructing or creating it. We can have a crucial influence on it, but it is in no sense logically related to us or our practices. An anti-realist concept of nature is in effect a contradiction in terms. Environmentalists dare not oppose science under the banner of some form of anti-realism. Because science claims to deal with reality, it may seem a promising strategy to challenge the former by attacking the latter. This will always be mistaken, and never more so when done in the name of an alternative vision. Visions that cannot themselves be grounded in reality are mere dreams.

We can only do science because of the nature of the world. We must not only be realists, but we have to suppose that the physical world contains a certain order and regularity. Science and reason cannot be identified because science itself depends on a rational understanding of the world. Many contemporary philosophers hold that there must be some 'neutral' ground on which we can stand, or a God's-eye view to aspire to, if something like the justification of science is to be attempted. We must somehow be able to step beyond our ordinary human rationality in order to judge its efficiency. How far reason, and the idea of reality, can be upheld in the face of such criticism is one topic of this book. Science needs legitimation, since the only alternative is a stultifying relativism, according to which the sciences, or even particular scientific theories, set their own standards. The snag then is that there is no point in adopting any of them. In many ways, science is a test case for rationality. If it cannot survive here, it will survive nowhere. For many, science is still the paradigm of human rationality, and must be concerned with the character of reality. Yet it has to be shown how this is possible. Even the physical sciences cannot be taken for granted. They rest on metaphysical foundations which must be made explicit.

This book builds on work I have done in previous books to defend the interlocked questions of rationality and reality against the onslaught of relativism and anti-realism. In 1973 my *Reason and Commitment* attacked relativism in science and elsewhere. I was particularly concerned in it with the influence of the later Wittgenstein and of T. S. Kuhn. Since then I have argued for a metaphysical realism which safeguards reality as a goal for all our intellectual endeavour, and particularly that in the sciences. That was the theme of *Reality at Risk* (second edition 1989). Over the last twenty years, relativism has become ever more respectable philosophically. In particular, attacks on the possibility of a reason which can even partially detach itself from its historical and social context have become ever more virulent. I have written about the place of rationality in society and in the social sciences in *Understanding Social Science* (1985).

It is common now to see science as a social practice, or as a language game, as something 'we' do (whoever 'we' are) and which cannot be given any further rational justification. This is all part of the repudiation of human reason as a guide to truth, and hence of the possibility of any kind of metaphysics. When confronted with a flourishing, even dominant, activity like science, this may not seem very threatening. Yet it removes the possibility of giving any kind of justification for continuing with the practice. Attacks on rationality put in jeopardy the very activities that are often being taken for granted. Without an ability to see what is true and to separate it from what is false, we cannot continue to practise the physical sciences as they have been traditionally understood. Our reasoning has to be rooted in the character of the world, whatever it may be. Otherwise the idea that science can explain anything, let alone everything, is sheer illusion.

1
Science and Reason

The Dominance of Science

Can science explain everything? Many assume that the only possible kind of explanation must be a scientific one, couched in causal terms. They see modern science, with its distinctive empirical method, as the exemplar of rationality. The undoubted success of the physical sciences in enabling us to predict and control much of our environment has persuaded us, it seems, that success in reasoning can only be obtained through science. There are, it appears, public standards of truth and falsity in science which in principle enable agreement to be reached. This seems far from the case in other aspects of human activity. Deep seated and, indeed, apparently irresolvable disagreement seems to be the norm once we depart from accepted procedures of checking, testing and proving. The position is not very good in would-be sciences such as economics. The inability of economists to give accurate predictions even of the effects of budgetary measures is notorious. Once we move into more controversial areas such as politics, morality and religion, notions of prediction and control are totally inappropriate. It is easy to conclude that there can be no tests of truth and hence no place for rationality in such areas.

The dominance of science in our culture is undoubted, and for much of the twentieth century its dominance in the field of philosophy has been complete. This may seem paradoxical as philosophy itself is far from being an empirical discipline. Yet for many philosophers the function of their subject has been to act as a handmaid to science. No longer has philosophy, let alone theology, been able to aspire to the title of 'queen of the sciences'. If the only path to truth is that of empirical discovery, it follows that philosophers sitting in their arm-chairs are unable to contribute anything of substance. Unless they get among the test-tubes of a laboratory, or the expensive gadgets of modern particle physics, it seems that they have no right to tell anyone anything about the world. All they have is their reason, and if human reason only works properly when schooled in the empirical methods of science, philosophical reasoning either has to be harnessed to the discoveries of science, or it is of illusory importance.

Things were not always so. The philosophical discipline of metaphysics often appeared to provide the basis on which all human intellectual enquiry, including empirical investigation, finally rested. This is a project dating from the days of Plato and Aristotle. Indeed the word 'metaphysics' was given by the editors of Aristotle's work to the books which came 'after the books on nature' (*meta ta physika*). The content and method of those books then became the exemplar of a whole way of thinking. Metaphysics has always been particularly concerned with what there is, with the nature of reality. That problem lies at the root of all questions, and it follows that metaphysical problems are the most fundamental and comprehensive that we face. We have to deal with the most basic presuppositions of our thought, which, just because they are so basic and pervasive, may easily be taken for granted and be totally unexamined. We may not even be conscious of what we are assuming. Yet because the issues are so central, they profoundly affect the way we live. It is not surprising that, despite its abstract nature, metaphysics has always been a battle-field on which each person's most cherished beliefs may seem ultimately to be at stake.

The entry on 'the nature of metaphysics' in *The Encyclopaedia of Philosophy* (vol. 5, 1967, p. 300) reads: 'Almost everything in metaphysics is controversial, and it is therefore not surprising that there is little agreement among those who call themselves metaphysicians about what precisely it is that they are attempting.'

In view of this, it is perhaps not surprising that many have taken refuge in the apparent certainties of science. When the practitioners of a discipline cannot even agree about what their subject is, it may not be surprising that 'metaphysics' can sink into becoming a word of abuse, applied to anything that is vacuous and incapable of proof. Much the same thing has happened to the idea of 'theology', which on some politicians' lips has come to mean anything that is too abstruse to be related to everyday life. Such an attitude of mind has in a sense been institutionalized in philosophy through the influence of the Vienna Circle. Its members met regularly in Vienna from 1925 to 1936 and propagated what they termed 'the scientific conception of the world'. Composed of philosophically inclined scientists and others, the Circle believed that everything had to be able to be reduced to the simplest statements about the 'empirically given'. As a result, their manifesto crisply states that 'the scientific world-conception rejects metaphysical philosophy' (Neurath and Cohen, 1973, p. 309). One example of a prime metaphysical statement is 'there is a God' and this would be rejected as 'devoid of meaning' because it cannot be reduced to a statement about what can be experienced. Knowledge, they would hold, is derivable only from experience, and 'this sets the limits for the content of legitimate science'. The Circle was not, however, only making a target of theological statements, nor did it consider metaphysics and theology to be synonymous. They may have thought metaphysical statements a 'residue' of theology, but all global statements referring to reality in general were treated with suspicion. Otto Neurath, one of the foremost members of the Circle, and inventor of its name, characterized the 'modern scientific conception of the world' as 'the interconnection of empirical individual facts, with systematic testing by experiment' (Neurath, 1983, p. 42). The idea was, with the help of logic, to create a unified science. The importance of experience in building up a scientific picture meant that sweeping generalizations, going far beyond the bounds of the empirically accessible, had to be ruled out. He said:

Theological residues in science can be suspected wherever empirical statements are related to a postulated or 'complete' insight
... The determinism of Laplace's formulation is untenable, for
the assumption of knowledge of an unlimited cross-section of the
world is totally meaningless. (p. 42)

Laplace had believed in a global determinism and thought
that only our ignorance of initial conditions makes successful
prediction and complete knowledge of the world unattainable.
Such an assertion is characteristically metaphysical in that it goes
far beyond anything we can experience and yet is comprehensive
in its scope. The Vienna Circle's reflex reaction to such claims
would always be to dismiss them as untranslatable into experience and hence as literally meaningless. They would always start
with specific experience, so that the general is logically connected
to the particular, and comes after it. Neurath himself (1983, p.
47) explicitly contrasts this approach with the traditional methods of philosophy and metaphysics, which would start with the
general. He echoes the Circle manifesto when he proclaims:
'Wherever there is a clear question, there is also a clear answer:
it makes no sense to speak of unsolvable riddles.' This is not as
reassuring as it sounds. If you can only ask questions which can
be answered, there can be nothing which we cannot solve. This,
though, is only because all other questions are ruled out of order
as not being clear, or having proper empirical content.

The notion of experience to which this view of science appealed
was one which depended on a view of raw, uninterpreted sensation
presented to us by the external world. These 'sense-data' were
supposed to form the foundation of all our knowledge. Colour
patches, sounds, smells and touch, were all combined so that we
could build up a view of the world. As the Circle's manifesto
claimed, 'something is "real" through being incorporated into
the total structure of experience' (Neurath and Cohen, 1973,
p. 308). With this statement, the Circle's members explicitly
denounced rival metaphysical views about the status of the real
world. Realism holds that reality is independent of human conceptions and the human mind. Idealism, on the other hand,

typically stresses the logical dependence of what is real on our minds. A simple illustration comes from our reaction to the question whether a clock ticks when no one hears it. If we hesitate about the way to reply, idealism, as a theory, can begin to gain a grip. Both realism and idealism, however, are global doctrines about the status of reality, and, as such, were dismissed by the Circle. The verdict was that, like all metaphysical claims, they were meaningless because without empirical content, and hence unverifiable.

The very nature of an empiricist approach, which begins with the experience of individuals, means that the resulting view of the world must be anthropocentric. What is beyond the reach of human beings can be safely dismissed. This is diagrammatically opposed to a metaphysics which starts from a conception of reality and then locates humanity in it. Part of what is at issue is the major difference in method. The Circle built their philosophy on the success of scientific method, allied to logic. They were thus 'logical empiricists', and were vehemently opposed to any idea that thinking on its own could lead to knowledge. Knowledge was gained through experience, they felt, and logical investigation proved that new information could not be produced by thought and inference. It is possible to deduce one set of statements from another, but everything in the one had already to be in the other. Logic or reasoning deals in tautologies, while our senses receive the main material we need for our understanding. This was the classic division between synthetic, or empirical, statements and analytic, or tautologous, ones, such as that all bachelors are unmarried. A major effect of making these linguistic distinctions was that it left no room for metaphysics.

The views of the Vienna Circle were popularized in English by A. J. Ayer, who was a participant in it. He summed up the position as follows:

We may accordingly define a metaphysical sentence as a sentence which purports to express a genuine proposition, but does, in fact, express neither a tautology nor an empirical hypothesis. And as tautologies and empirical hypotheses form the entire class of significant propositions, we are justified in concluding that all metaphysical assertions are nonsensical. (Ayer, 1946, p. 41)

Such an outlook appears to give science firm foundations. The demand for verification and falsification by actual and possible experience does indeed lead to the epistemological doctrine of foundationalism, according to which our knowledge can be based on an incorrigible and certain starting-point. Reason is tamed and put in the service of the information which we passively receive. It can help us to systematize our experience, but cannot transcend it. Philosophy may help in the task of clarifying scientific concepts, but cannot legislate for science. There is no ground left for it to stand on. The downgrading of the power of human reason thus goes hand in hand with a diminished view of philosophy. The dismissal of metaphysics as meaningless does not just result in giving up high-sounding phrases with no 'cash-value in the real world'. It also means that we cannot reflect on the nature of science. In particular we cannot justify it.

Why should anyone do science in the first place? Why should we adopt 'scientific method', if there is such a thing? A natural answer to these questions is that science tells us the truth, and, indeed, science is the source of truth, if we take the Vienna Circle's logical positivism as our guide. This approach can be termed 'scientific', but it is still in need of justification. We may be impressed by Western science simply because the technology derived from it literally delivers the goods. Without it we could not have colour television, aircraft and all the gadgets of modern life. A cynic might add that without it we would not have nuclear weapons, missiles and the gadgets of modern warfare. There is no longer the assurance that science can be linked to the idea of inevitable progress. The scientific knowledge that produced the gas ovens of concentration camps or the bombs that devastated Hiroshima and Nagasaki cannot be unequivocally welcomed. The world at the end of the twentieth century is not obviously an improvement on the world at the beginning. Every medical advance can be counterbalanced by some more clever method discovered by science of inflicting suffering on humanity.

Difficulties for Empiricism

We are bound to be affected by the general doubts arising in our culture about the worth of science. This makes it all the more dangerous if we take science for granted and only allow philosophy to reflect on its findings. The question why we should rely on science, and particularly why we should rely only on it, arises naturally from the state of the world we live in. When the Vienna Circle ended their manifesto with the ringing proclamation that 'the scientific world-conception serves life' (Neurath and Cohen, p. 318), they were saying someting that perhaps rang true in 1929. Yet Wittgenstein, who was at first received by the Vienna Circle as one of their own, commented in 1947 that it was not in the least absurd to believe the age of science 'is the beginning of the end of humanity' (1980, p. 56). He was explicitly willing to entertain the possibility that 'the idea of great progress is a delusion, along with the idea that the truth will ultimately be known'.

The bitter experiences of the twentieth century have predisposed many to distrust science. Far from being the engine of progress, it can appear to be an instrument of oppression. The air we breathe and our rivers and seas become more polluted as a result of the very technology which was intended to transform our world for the better. An intellectual climate is arising which is unwilling to take the products of science on trust. Science is forced to justify itself. What justification for science did empiricists and positivists produce? They allowed physical science no metaphysical foundation. That did not worry them as they believed that metaphysics was nonsense, but it does mean that science can claim no rational basis. Reason is left with no scope for standing outside science to justify it. Logical positivism was in effect an attack on the possibility of human reason outside the strict constraints of deductive logic. It had distinguished forbears such as Hume, but he too was unable to give any justification for the practice of science.

The most glaring example of the undermining of rationality is provided by the verification principle itself. We can certainly rule out a great many statements as nonsensical if we make our ability to verify a proposition the criterion of its being meaningful. It has, however, been notorious that the verification principle itself cannot be verified. The starting-point of logical positivism cannot itself be justified and indeed by its own lights should be regarded as meaningless. Ayer claimed that the principle could be regarded as an axiom, but this fails to meet the challenge of why we should adopt such an axiom. The very fact that it stops us saying what we might wish to say suggests that it is a controversial criterion of meaning. This problem of reflexivity arises in dramatic form here. The principle when applied to itself, as it must be, undermines itself. It cannot meet its own demands. Sawing off the branch one is sitting on is not generally regarded as good practice in human life, and such damaging reflexivity must always be seen as a warning that something is going wrong with our reasoning.

The Vienna Circle also found difficulties in justifying the practice of science, which typically searches for regularities in the physical world and tries to discover order. Science would be impossible if everything were absolutely chaotic. Indeed, as we shall see, one of the fascinating aspects of contemporary science is the way it shows how chaos is itself ordered. No two snowflakes may be identical but that does not make the behaviour of snow unpredictable. This problem of order in the physical world, and the question of the uniformity of nature, goes hand in hand with that of the comprehensibility of nature. One solution is to indulge in metaphysics and give a rational explanation why the world is intelligible and orderly. Another is to deny that the solution lies in the world at all, or alternatively to say that there is no solution. Hume took the latter course and pointed out that there could be no demonstrative argument to prove 'that those instances, of which we have had no experience, resemble those of which we have had experience'. His diagnosis was that our supposition that the future resembles the past is 'derived entirely from habit.' (1888, pp. 89, 134). For Hume, human custom replaced the possibility of rational justification. Kant, on the other hand, was much more willing to embrace metaphysics but claimed that it 'is taken from the essential nature of the thinking faculty' (1970, p.

9). In other words, we order the world not because of its inherent nature, but because that is how we naturally see things. The ordering is a product of our intellect, not of reality. Kant thus introduced the idea of a 'synthetic a priori' principle, which organised our experience and was not derived from it. He claimed that 'the understanding does not draw its laws *a priori* from nature, but prescribes them to nature' (1955, p. 82).

This idea of *a priori* synthetic judgements, which demolished the basic empiricist idea that the physical world provides us with all the information we need, was, needless to say, resisted by the Vienna Circle. Its members argued that we do not obtain knowledge through the impress of human reason on informed material, perhaps like the stamp of a signet ring in wax (Neurath and Cohen, p. 312). They were sure that the material itself was ordered in a certain way, but escaped the need for metaphysical explanation by taking the empiricist view that such order cannot be known beforehand. They said: 'The world might be ordered more strictly than it is: but it might equally be ordered much less without jeopardizing the possibility of knowledge.' The Circle accepted that induction, defined as 'the inference from yesterday to tomorrow, from here to there', is only valid if there is regularity. Their position is that regularity does not have to be presupposed in some *a priori* way, but can be derived empirically. Science thus may depend on regularity, and will itself discover it. It pulls itself up by its own boot-straps. If we do not discover much regularity, our science cannot progress.

It is taken as obvious that we should pursue science as our sole source of knowledge. If the world proves sufficiently ordered for us to find it intelligible, that appears to be a lucky chance. It is, however, the chance on which our whole system of knowledge rests. We are told that the method of induction may be applied 'wherever it leads to fruitful results, whether or not it be adequately founded' (Neurath and Cohen, p. 313). The important point is that all inductive inferences have to be tested empirically. A positivist would hold that there could be no other kind of test. The notion, however, of obtaining fruitful results from a method that is not properly founded is curious. It seems that every inductive inference is a leap in the dark, which is justified only if we land safely. As a way of giving us confidence to leap, this has it flaws. One difficulty is recognizing the difference between results that are fruitful as opposed to those that only appear so. We may think we are discovering regularity when all we find is a series of coincidences. The latter indeed is perhaps more likely if we have no right to expect regularity. Since Hume, empiricists have found it very difficult to distinguish patterns of phenomena which merely happen to appear together from those which possess some causal link between them. Regular coincidences merge into causally linked events for an empiricist, since there is nothing other than apparent regularity for experience to discover. The idea of underlying causal connections as an explanation of regularity is not an empiricist one.

The perennial problem with testing induction empirically is that even when predictions prove successful this will only support the principle of induction if we already accept it. We still face the question why the success of past predictions should be a guide to the future. There is an endemic circularity here. We may wish to appeal to previous empirical successes as a guarantee of the regularity of the world, and hence as the justification of induction. Yet that presupposes that the apparent regularity will continue, which is the point at issue. Empiricists must predict future experience without any metaphysical underpinning. No appeals to the nature of reality and the fact of physical causation can be consistent with the empiricists' outlook. The latter deliberately leaves reason with little scope, and any notion of justification must fail to get a grip. Experience is everything.

Wittgenstein's Attack on Metaphysics

The Vienna Circle's manifesto concluded with a list of names which included that of Ludwig Wittgenstein, who was identified as a 'leading representation of the scientific world-conception'. Yet it is a gross simplification to suppose that his early philosophy was narrowly positivist. In the *Tractatus* he sums up what he saw as the correct method in philosophy:

To say nothing except what can be said, i.e. propositions of natural science . . . and then, whenever someone else wanted to say something metaphysical, to demonstrate to him that he had failed to give a meaning to certain signs in his propositions. (1916, 6.53)

This may look like the Vienna Circle's condemnation of metaphysics, but there is a catch. Wittgenstein had pointed out in his previous section (6.523) that there are things that cannot be put into words. He continues: 'They make themselves manifest. They are what is mystical.' He certainly thought that what can be said must be said clearly and 'what we cannot speak about, we must consign to silence' (7). There is much to indicate that Wittgenstein thought that what lay beyond language, so far from being nonsensical and of no importance, was actually more important than what could be confined within its limits. Yet in stressing the limits of language, he also meant to limit the power of reason to get to grips with issues such as the justification of science.

Wittgenstein in the *Tractatus* did explicitly deal with the problem of the principles of science on which its methods depend. Questions about causality and induction are presupposed by science, he thought, and could not be a matter of empirical discovery. They are, he believed insights 'about the forms in which the propositions of science can be cast' (6.341). His problem was that if science did not discover such truths as facts about the world, and if they were not necessary truths or laws of logic, it was difficult to see a third way. He had abandoned metaphysics and accepted that metaphysical statements could not claim meaning. He used an analogy to elucidate the position, suggesting that we imagine a white surface with irregular black spots on it. By covering it with a fine, square mesh, we could say of every square whether it was white or black, thus improving 'a unified form on the description of the surface' (6.341). It is perhaps significant that he insists that the form is optional. We could have used different kinds of mesh, and he points out that 'it might be that we could describe the surface more accurately with a coarse triangular mesh'. The presuppositions of science are like the mesh or net, telling us nothing about the world but enabling us to describe it.

One might seize on the term 'optional' and think that Wittgenstein contends that it does not matter which set of principles we adopt to enable us to do science. The choice would then appear arbitrary or conventional. In the *Tractatus*, however, with its picture theory of meaning, Wittgenstein was concerned with the relation of words to the world; he thought that truth mattered. Wittgenstein's own example was Newtonian mechanics, and he believed that scientific schema was an attempt to construct according to a single plan all the *true* propositions we need to describe the world. Different nets will divide up the world differently and it is important how this is done. Nevertheless, they are not themselves part of natural science, and do not themselves tell us about the nature of the world. What they picture is the formal structure of reality, and not any particular fact.

Wittgenstein's problem is the status of the presuppositions of science, given that he has ruled out the meaningfulness of metaphysics. As a result, we cannot give reasons for such principles. Science cannot justify its own assumptions and we cannot step back from say physics to metaphysics in order to give physics a rational foundation. The scientific assumption that there are laws of nature, so that things occur in a regular manner, is something that, according to Wittgenstein, 'cannot be said' (6.36). Instead, he says, 'it makes itself manifest' or shows itself. A structural property, like logical form, is only indicated in the fact of something being structured. Wittgenstein uses the same phrase about the presuppositions of science making themselves manifest, as when he refers to the 'mystical', which shows itself but cannot be put into words. Whatever he meant to include in that category, his restriction of meaningful language to the subject-matter of science is extraordinary. Not only does it rule out much of importance that may lie beyond science, but it robs science of a rationale that can be explained and justified within the confines of language. Too much is consigned to the category of the ineffable.

The word 'mystical' is important for Wittgenstein, as can be illustrated by his use of it when he said: 'It is not *how* things are in the world that is mystical, but *that* it exists' (6.44). Nothing could demonstrate more his feeling of the inadequacy of physical science as a total explanation of the world. That things exist at all is a fact which science cannot explain. While Wittgenstein appears to agree with the logical positivists about the meaninglessness of such cosmic questions, he also paradoxically felt that the questions mattered. Science can in fact only be relied on to explain everything if there is an arbitrary restriction on what can count as real and what is to count as a proper explanation. The continuing problem is why such restrictions should be made. The price of saying there are no unsolvable riddles left is to limit severely what can count as a genuine riddle. Only what is accessible to science will be counted as real. Yet this merely shifts the issue to the question of the scope of science. Do we mean, for example, science as it happens now to be, or science as it might one day develop? The positivists recognized that they had to talk of *possible* sense-experience, and this made it possible for Ayer in 1936 to accept the meaningfulness of the proposition that there are mountains on the farther side of the moon (1946, p. 36). In fact this could be checked some thirty years later. Many propositions, even within science, cannot even in principle be checked in such a direct way. Are claims about the other side of the universe to be ruled out as unscientific? Scientific theory increases its scope year by year. The meaningfulness of theories about the far reaches of the universe, or, at the other extreme, sub-atomic particles, cannot be linked too closely with the possibility of human observers gaining access to them. Indeed as the importance of theory has been stressed more and more in the philosophy of science, it has been recognized, even by those influenced by empiricism, that a straightforward cashing out of theory in terms of its empiricial relevance is too simple. There cannot be a close correspondence between theoretical terms and empirical observations. The idea of unobservable entities is no longer an embarrassment as it was for the verificationists.

The idea that science is the source of all explanation runs deep in the modern world. Wittgenstein shows the distance between himself and the Vienna Circle when he expresses reservations about the scope of causal explanations. He clearly means to be critical of modernity when he comments that 'the modern system tries to make it look as if *everything* were explained' (1961,

6.372). Modernity and science certainly sit very happily together. Indeed confidence in science as the arbiter and exemplar of human rationality lies at the very heart of the modern world. Only in recent years has this confidence begun to crack.

The *Tractatus* accepted that the presuppositions of science were neither logical truths nor empiricial discoveries. Their status seems somewhat anomalous. When Wittgenstein changed his whole outlook, he was still faced with the same problem. He no longer believed that language functioned in only one way, by picturing reality. The propositions of natural science could no longer claim a monopoly of meaning. Instead he came to believe that the meaning of a word is the way it is used. Language can be used in a multiplicity of ways and meaning depends on context (Trigg, 1991). We are often tempted to transfer the meaning of a word in one context to its use in another. Our failure to recognize the different uses suggests to Wittgenstein a source of the philosophical confusion he wished to lay bare. Metaphysics was a particular target, since he thought it arose simply because of an insistence on using language appropriate to one context in a totally different one.

It is intriguing that despite the latter Wittgenstein's radically different approach to the question of meaning from that of a positivist, metaphysics is still anathema. For example he lists words such as 'knowledge', 'being', 'I' or 'object' which characteristically appear in the writings of those who deal in metaphysics (1953, 116). His sharp question to philosophers using them is: 'Is the word actually ever used in this way in the language-game which is its original home?' This comparison with games is meant to invoke their rule-governed character. Wittgenstein's point would be that a word like 'object' gains it meaning from the context in which it is generally learned. It has no special meaning over and above what a child could grasp. Yet once it is torn from the everyday world and takes on a rarefied usage, we run the risk of talking nonsense without realizing it. We have ignored the rules.

Wittgenstein therefore claims: 'What *we* do is to bring words back from their metaphysical to their everyday use.' He is here explicitly attacking the language and assumptions of the *Tractatus*. For example, he refers to his statement in that book of the general form of propositions as being 'This is how things are'. He has become dissatisfied with such sweeping generalities and says, in an analogy reminiscent of the net: 'One thinks that one is tracing the outline of the thing's nature over and over again, and one is merely tracing round the frame through which we look at it' (1953, 114). Yet what makes us choose one frame rather than another? We are left with the problem that if metaphysics is merely the science of language, we seem to have no resources for rationally justifying our most basic stances.

When Wittgenstein talks of the necessity of returning metaphysical language to its everyday use, it is hardly surprising that he regards his own *Tractatus* as badly in need of such a return. In it he had, for example, claimed: 'Generally speaking, objects are colourless' (1916, 2.0232). To make sense at all, that has to be understood as use of language in a highly technical way. The later Wittgenstein is content with 'the language of everyday' (1953, 120). Philosophy cannot then legislate about the use of language, but should only describe what is said. Philosophy cannot give language any foundation. He says: 'It leaves everything as it is'. Once again we arrive at the conclusion that the possibility of rational justification is illusory. Instead, our reasoning finds its place in the way we live. It cannot, he holds, abstract itself from the situation in which we find ourselves and function outside all contexts in a complete vacuum. Yet that is what metaphysics would have us do. A traditional metaphysical view is one of the self reasoning about truth in a manner that can be detached from place and time. Since for Wittgenstein the context is the source of meaning, he finds this an impossible picture. Language will lose all claim to meaning once it is abstracted from its normal role in the activities of our life.

Wittgenstein made much of the term 'language-game' and through it he tried to stress the intimate connection between language and the way we live. He says: 'The speaking of a language is part of an activity or of a form of life' (1953, 23). Our understanding has to be rooted in particular practices. The idea of a detached reason establishing, however tentatively, what had to be true of the world, was anathema to him. Instead, he emphasized 'grammatical' issues about the nature of concepts. In effect, he was unwilling to allow philosophy, with all its deficiencies, to try to get to grips with the real world. Yet it can be fairly pointed out that this charge presupposes that the words 'real world' have a determinate meaning. Wittgenstein pursues this point with vigour, arguing that our very concept of reality will be rooted in the form of life in which we are situated. He still uses the word 'picture' which he had used in the *Tractatus* to link with reality. He had held there that 'a proposition can be true or false only in virtue of being a picture of reality' (1961, 4.06). Yet in the last year or so of his life he was saying: 'I do not get my picture of the world by satisfying myself of its correctness' (1969, 94). Instead, he claims 'it is the inherited background against which I distinguish between true and false'. What is the status of this kind of framework which gives me my standards of truth? This is not such a very different question from that about the status of the net. We are still searching for justification.

The later Wittgenstein seems willing to subordinate questions of truth, and of what counts as a good reason, to particular conceptual systems. (See my *Reason and Commitment* for a discussion of Wittgenstein and relativism.) In this, he lays himself open to the charge of relativism. Truth is no longer, it seems, to be associated directly with the world. In fact he does not hesitate to say that description of a world-picture could be part of 'a kind of mythology' (1969, 95). Their role is to be like the rules of a game. This conception of rules is central to the thought of the later Wittgenstein. They are public and social in character, and the use of language will only make sense in public contexts. Yet faced with different, and possibly incompatible, language-games, we may well wonder why we should play this game rather than that, or indeed any game at all. If we pursue the analogy we may wonder what arguments could be presented to someone who fails to appreciate the game of cricket. Cricket can be described, but can one show others the error of their ways in not playing or watching it? Are they wrong? One can in fact do no more than explain that this is how the game is played, and that is precisely Wittgenstein's retort to anyone searching for justification, or for reasons and grounds, for taking part in a particular language-game. It seems that language-games as such are not true or false.

Their rules cannot be justified and they rest on no foundation. They cannot be 'tested' since what counts as an adequate test has to be a matter internal to the language-game (1969, 82). As Wittgenstein says: 'The *truth* of certain empirical propositions belongs to our frame of reference' (1969, 83). The frame may dictate how we see things, but, like the net, it can be displayed but not justified. There is nowhere else for us to stand in order for us to pass judgement.

The Removal of Reason

Reason and philosophy itself are put in jeopardy by the later Wittgenstein. In fact, so is any notion of reality. The *Tractatus* had a robust sense of reality, even though language was restricted in it to what was assessible to science. Once, however, everything becomes internal to language-games and forms of life, reference to reality or the world has to gain its meaning from a particular context in a particular human practice. When everything depends on our frame of reference and there is no way of testing or reasoning about the frame, our most fundamental beliefs about reality will have an arbitrary starting point. Even if our practices are based on certain general facts of human nature, these are irrelevant to the question of justification. As Wittgenstein points out, speculation about which facts of nature give rise to our concepts is a causal investigation and 'we are not doing natural science' (1953, xii). The bedrock of our concepts may lie in the 'natural history of human beings', but that lies beyond the scope of philosophical justification (1953, 415).

The later Wittgenstein was convinced that we cannot discuss how far one view or another may 'agree with reality'. He says that 'with this question you are already going round in a circle' (1969, 191). What one counts as real depends on our basic stance towards the world. Empiricists will say that experience will teach us what reality is, and to some extent the early Wittgenstein accepted this, by tying the meaningfulness of language to what was within the scope of science. There is still, though, the problem of justifying our reliance on experience. As he says at the end of his

life, 'experience does not direct us to derive anything from experience' (1969, 130). He says: 'If it is the *ground* of our judging
like this, and not just the cause, still we do not have a ground for
seeing this in turn as a ground.' All justification has to come to
an end. Otherwise we are involved in an infinite regress. This is
a recurring problem in any theory of rationality. We can have a
reason *y* for having a reason *z*, but we must then provide a reason
for adopting reason *y*. Do we retreat to reason *x* and so on for
ever? This is not a trivial point, and it leads many to believe that
we might as well stop the whole process of justification sooner
rather than later. He adopts a holistic approach, which stresses
the way in which beliefs form a system and give each other natural
support (1969, 142). Within the system, everything is coherent
and hangs together. The problem arises when the whole system
is put in question. As Wittgenstein remarks: 'The difficulty is
to realize the groundlessness of our believing' (1969, 166). This
means that he is unable to defend even those who trust contemporary physics rather than oracles (1969, 499). The later
Wittgenstein's whole approach is precarious precisely because of
its lack of any proper foundation. The Vienna Circle had claimed
to provide foundations for knowledge, but themselves failed
because of their repudiation of the grounding afforded by
metaphysics.

The basic principles of science certainly cannot be given a
rational justification by the later Wittgenstein. He finds that even
more problematic than in the *Tractatus*. He admits that 'the "law
of induction" can no more be *grounded* than certain particular
propositions concerning the material of experience' (1969, 499).
Yet he cannot say it is true. In the end, everything comes down
to human practices. Language-games are just there and depend
on some prior trust in something (1969, 9). Dealing with the
question of our knowledge of the boiling point of water, he con-
cedes that the behaviour of water could change in the future. He
thus for a moment implicitly allows that there is a world to which
language has to conform. Nevertheless he explains that 'we *know*
that up to now it has behaved *thus* in innumerable instances'. 'This
fact,' he continues, 'is fused into the foundations of our language-
games'. Yet any idea that reality could act as some kind of

constraint on our language-games is swiftly denied. There can be no external standard, providing a basis for justification. Once more he insists that our language-game is not based on grounds, nor can it be reasonable or unreasonable (1969, 559). He says of it, 'It is there – like our life.'

Wittgenstein links the idea of certainty with his notion of a form of life (1969, 358). It, too, lies beyond being justified or unjustified, but is 'something animal'. We just do naturally expect certain things and live in a particular way. Both Wittgenstein and Hume, having rejected the possibility of metaphysical justification, have to turn back to appeal to human nature, or something very much like it. Metaphysics is replaced by brute facts about human existence. Some may wish to explain such facts by turning to neo-Darwinian doctrines of evolution and natural selection, but this is in no way a justification for our behaviour. As we shall see later, it is a causal explanation and entirely different from giving reasons. Whether reasons are forthcoming is an important question, but it is not one that can be answered by changing the subject.

The division by the Vienna Circle between logic and empirical statements was a simple way of disposing of metaphysics. Wittgenstein, in his later period, proved to be no more in love with metaphysics than earlier. He did recognize, though, that there could be no sharp boundary between logic and empirical propositions (1969, 319). He enlarges on this by saying that 'the lack of sharpness is that of the boundary between rule and empirical proposition'. Rules constitute language-games and determine what we take to be evidence. The rules that are implicit in our practices govern what we consider reasonable and unreasonable, but cannot themselves be rationally defended. The principles of science are rules of the scientific language-game. The game may change over time, but cannot be grounded.

The notions of language-game and form of life are central to Wittgenstein's later philosophy but are hard to define clearly. Indeed his philosophical method encouraged him to resist defining the 'essences' of things, and he was much happier with the idea of family resemblance. One of the clearest messages to come from the later Wittgenstein was his opposition to the possibility of metaphysics and the idea of a free-floating reason, transcending its local circumstances. His earlier view of the principles of science as a net became transmuted into the idea of rules governing different language-games. Our rational understanding can only be seen as rooted in our social practices. Indeed there may even be a faint echo of Nietzsche in his attack on the very possibility of reason (see Trigg, 1988, chapter 10). His emphasis on the nature of concepts, and on the social setting which gave them their meaning, left no way in which philosophy, in general, and metaphysics in particular, could get to grips with the real world.

All this has a shattering impact on the status of science. It can no longer itself be seen as an arbiter of truth, but becomes one system amongst others. It can only be justified in its own terms, but there can be no rational basis left for upholding it in the face of opposition. We may go on being scientists because that is the way we have been educated, but faced with alternative societies preferring to put their trust in oracles, astrology or whatever, we can only indulge in name-calling. We may not expect to convert or persuade those from a different background, but more disturbing is the fact that, once challenged, scientists have no means of justifying even to themselves the practice of science. It is what they do, and that is all. Perhaps this is not a genuine criticism if such justification is in principle impossible. It seems unlikely, however, that the questions why we should practise science at all or trust its pronouncements, perhaps to the exclusion of other views to which we may be attracted, can be shrugged off so easily.

The Vienna Circle would have been horrified by the way in which Wittgenstein's later philosophy allows religion back on the scene. It cannot, it seems, claim objective truth, whatever that may be, but as an undoubted human practice it generates its own rule-governed activities and its own meaning. It is deprived of any metaphysical grounding, and many religious believers could feel that this inability to claim truth itself undermines religion. However there is also the point that in such circumstances religious practices can claim to be meaningful and cannot be shown to be false. D. Z. Phillips, for example, has systematically applied Wittgensteinian views to Christianity. He maintains that 'The

meaning of what agreement to reality comes to is itself determined by the language-games we play and the forms of life they
enter into' (1988, p. 55). His point is that we cannot talk of a
reality external to our beliefs and practices. Language cannot be
viewed as a screen which may hide God (p. 289). Opposing the
view (expressed in my *Reason and Commitment*) that belief in God
is distinct from the commitment that may follow it, and is the
justification for it, Phillips quotes with approval a remark by Nor-
man Malcolm that this desire for justification, the desire to ground
religious belief in some kind of ontology, is 'one of the primary
pathologies of philosophy' (p. 267).

The issue is once again the meaningfulness of metaphysics,
just as it was in logical positivism. Yet the question of the very
intelligibility of the concept of objective reality marks a radical
departure from the limited certainties of empiricism. The argument has become one about the possibility of transcendence,
and not just the transcendence of God. The existence of anything in and by itself, apart from our practices, appears to be put
into question. Indeed at times the later Wittgenstein has been
accused of linguistic idealism. Yet without an ontological anchor
not only religious practices are cut free from their moorings.
Each human way of life has to be accepted in its own terms, since
there is nothing against which they can be measured. No language-
game can make global claims to truth. The very idea of the
rational justification of whole practices is dismissed as an impossibility. Human reason, except in its local manifestations, is left
with no role. Our thinking is too firmly rooted, it is alleged, in
particular linguistic practices.

One may indeed idly wonder why metaphysics, as well as physics,
could not count as a rule-governed language-game. Wittgenstein
could presumably deny any connection between metaphysics and
ordinary human life. The concepts of metaphysics, it would be
alleged, are too far removed from the contexts in which they
could be given any application. Physics rests on genuine expectations about the regularity of physical events. The question is
where metaphysics could gain a hold. Perhaps it does answer
genuine human needs, as we reflect on the apparent contingency of things, and on our own place in the world. Wittgenstein

would have none of this, however, and his willingness to dis-
qualify some ways of talking as genuine language-games serves
to raise again the issue of what is to count as one. The matter
is graphically illustrated by D. Z. Phillips' treatment of religion.
According to him, there can be no way of rationally resolving the
dispute between theists and atheists. Yet he goes further and
accepts that even within Christian theology there are warring
conceptions which reflect 'deep religious differences' which cannot be rationally resolved (1988, p. 240).

In such contexts, it seems, appeals to truth and falsity, and to
such notions as heresy, are mere slogans. All we may be left with
is the fact of lack of agreement. Even Christianity, then, cannot
be regarded as one form of life, but seems to be composed of an
indefinite number of overlapping ones. Indeed, if insoluble
disagreement is the sign of a clash of forms of life, we may each
in the end find we belong to a form of life with one member,
and that is emphatically not what Wittgenstein meant. At times
he appears to be claiming that human life as a whole constitutes
a form of life. The ideal of a form of life cannot in fact be given
a clear content. Any account of human activity is liable to lapse
into incoherence without such notions as reason, truth and
reality. Certainly without them, all human belief, and not just
religious belief, will lose its point. Science is as much at risk as
any religion.

2
Science and Pragmatism

'Science Works'

Any discussion about the possible justification of science is likely to be met by the conviction that whatever its rational foundations, it certainly works. Scientists seem progressively more able to control the physical world. It has made more difference to the lives of ordinary people than any other human activity. Modern life is dominated, even cocooned, by the products of this century's physical science. Every time we turn a switch to obtain light or heat, every time we watch television or get a cool drink out of the refrigerator, we affirm the obvious success of science. Our transport by road, rail, air or sea depends on modern inventions. Life, in short, would be unimaginable without the benefits of modern science. What need do we have of philosophic theories which claim to give an underpinning to scientific theory? Science works.

This might seem to be the reaction of the ordinary person whose feet are planted firmly on the ground, the gut feeling of the individual called in a slightly earlier age 'the man on the Clapham omnibus'. Yet it is also the starting point for a sophisticated philosophic position, pragmatism, which has flourished primarily in the United States since the end of the nineteenth century. Philosophers such as William James, C. S. Peirce, and John Dewey, have helped to mould a distinctive outlook. According to Peirce, the maxim of pragmatism is that one conception could have no difference in 'logical effect or input' from another, except in so far as 'it might conceivably modify our practical conduct differently from that of our second conception' (1957, p. 252). The whole weight of attention should be on the way in which people's behaviour is modified. For the pragmatists, as for Wittgenstein, metaphysical speculation should be ruled out. Instead of reasoning about the nature of reality, or searching for firm foundations of knowledge, the pragmatist typically believes that we should start from where we are and build up our conception of knowledge out of our present practices. Because metaphysics tends to point to a reality beyond our knowledge, it can open up the possibility of scepticism. If something is beyond our grasp, it might appear that we have no very great reason for believing it is there, and it is easy for anyone to deny its existence. Pragmatism begins with our actions and our actual purposes, and so avoids this.

This is demonstrated by the manner in which pragmatists define knowledge. John Dewey, for example, said: 'If we can see that knowing is not the act of an outside spectator, but of a participator inside the natural and social scene, then the true object of knowledge resides in the consequences of directed action' (1960, p. 221). He said later on: 'The criterion of knowledge lies in the method used to secure consequences and not in metaphysical conceptions of the nature of the real' (p. 221). In the case of science this has to mean that our scientific method is the source of knowledge. We construct knowledge on the basis of tried and tested procedures, so that it grows out of our practices. This involves a marked change from the Greek idea of knowledge as reflecting a separately existing reality and of contemplation of that reality as the highest good. The deep division between theory and practice which has pervaded philosophy ever since is allegedly demolished by pragmatism. Indeed Dewey himself rooted the distinction in the social divisions of ancient Athens. He pointed out the distinction between the leisured class of citizens to which philosophers naturally belonged, and the workers and non-citizens who had to do all the work:

The definitely socio-practical division . . . was converted by philosophical formulation into a division between practice and theory,
experience and reason. Strictly scientific–philosophic knowledge
and activity were finally conceived to be supra-social as well as
supra-empirical. They connected those who pursued them with
the divine and cut them off from their fellows. (1938, p. 73)

Whatever knowledge is, then, it must stem from our social
practices, and not be understood as referring to some transcendental, timeless reality. According to Dewey it is not about any
kind of independent reality, because he was opposed to the whole
idea of any reality which could somehow be viewed from the
outside. This was why he criticized what he termed 'the spectator
notion of knowledge' (1977, p. 90). The latter can in one way
seem a very natural way of conceiving things. If I know, or even
believe that it is raining, it is not ridiculous to imagine that there
is a distinction here between the subject and object of the
knowledge or belief. It may be raining whether or not I am aware
if it. I can be totally separate from what I have beliefs about,
although I am myself a part of the same real world. Sadly, my
beliefs may often not coincide with what is actually happening.
Yet the pragmatist would resist this sundering of subject and
object. Dewey could not accept the realist idea that reality is
independent of our conceptions of it. He ruled out any general
theory of reality. Our understanding of what is real comes from
what we do. He says: 'Pragmatism is content to take its stand
with science. It also takes its stand with daily life' (1977, p. 88).
Pragmatists object to metaphysics because it fails to guide our
actual practices. Instead of starting with the conception of some
reality which may not even be attainable, we must, they believe,
begin with the concrete things around us in our every day world.
William James put the matter succinctly (1907, p. 45) when he
said that the pragmatist method aimed to settle metaphysical
disputes that might otherwise never end. It appeals to what, if
any, would be the practical consequences of the truth of one
theory, rather than another. In a famous image he said that 'you
must bring out of each word its practical cash value' (1907, p. 53).
    We should not be too hasty in concluding that metaphysics

makes no difference to what we do. The idea of a pragmatic justification of a metaphysical position is not self-contradictory. Indeed, one possible way of defending a realist approach in science could itself be based on a pragmatic approach to science. This would be unacceptable to Dewey, but it is often noticed that working scientists do tend to be realists. They think they are investigating a physical world that is independent of their own operations. Indeed this might even seem to be 'common-sense' in science. The very fact that people who believe this appear to get results might seem to be its own justification. Nicholas Rescher in fact argues for a metaphysical realism on this basis. He claims that 'Realism is not a factual discovery but a practical postulate justified by its utility or serviceability in the context of our aims and purposes' (1992, p. 270). His argument is that, although we do need to presuppose the objectivity of the world to make sense of the very idea of scientific inquiry, we are in a sense pulling ourselves up by our own bootstraps. We need the idea of a knowledge-transcending reality to give us the goal we need in the pursuit of new knowledge. He claims, in words that would be explicitly repudiated by many pragmatists, that 'inquiry, as we standardly conceive it, is predicated on the commitment to an inquiry-independent reality' (p. 263). Our whole impetus to inquiry is, he thinks, given by this belief in reality.

Whether there is a reality independent of the way we conceive it, and how we can refer to it, is the most central of all metaphysical questions. Metaphysical assertions about the character of reality are typically based on the presupposition that so-called 'reality' is not a human creation. What exists is a separate issue from any question about our beliefs or thoughts about it. Ontology should not be linked too closely with epistemology. At first sight, that seems to be what Rescher is saying. Yet although he appears to be underwriting a realist position, and he is making concessions to realism, the difference between pragmatism and realism is illustrated well in his approach. He starts with the fact of inquiry, and he questions what assumptions we need to conduct our science. His policy is not to start with questions about what is true. Instead he makes reality a Kantian 'thing-in-itself' or *noumenon*. We can aim at it, but will not reach it, and so we can only hope to give our 'best estimate' of what is actually true. Rescher is emphatic that our judgements of truth are fallible and revisable, and he parts company with Peirce's belief that scientists could ever attain complete truth (1992, pp. 47–8).

Rescher says: 'Our knowledge of the world must be presumed incomplete, incorrect and imperfect, with the consequence that "our reality" must be considered to afford an inadequate characterization of "reality itself"' (p. 262). Rescher is here denying that science could ever be complete, on the Kantian ground that reality is a regulative ideal, to which we always aspire, but which in principle we can never reach. An analogy, though it is not used by Rescher, would be that of goal-posts. The rules of football lose their sense without the overarching purpose of scoring a goal. Yet to pursue the analogy, for Rescher the game is one in which goals are never in fact scored. He thinks that nothing can be said about transcendental reality. It recedes from our grasp because we can never be in a position to make final and infallible claims to knowledge. Yet at the same time, our scientific practices would quickly lose their point without some conception of a reality to aim at. It is at this stage that the pragmatist flavour of Rescher's position becomes clearer. He says: 'Negative and regulative though the conception may be, we nevertheless require it as a tool of indispensable utility, for it alone enables us to operate our standard concept-scheme in matters relating to enquiry' (p. 264).

The idea of any transcendent reality is thus being used merely as a tool, in order to give sense to our practices. The mode of argument is essentially that science works, but that we have to make certain assumptions to practise it, chief of which is that we are investigating objective reality. If we cannot do science without this assumption, this seems to be some kind of justification for it. Yet the parallel argument about football would suggest that because we cannot play the game without goal-posts, we must have them. This may be so if we want to play an organized game of that character, but it does not begin to address the question why we should play the game in the first place. Why play any game at all? If we want to join in something, why not try cricket instead? It may seem curious to ask for a justification for playing games. If we enjoy doing so, that would seem to give purpose enough. In the case of science, however, enjoying it seems a far from adequate explanation of the practice. It may indeed be worth engaging in even if we do not enjoy it. Is it really enough to insist, with the later Wittgenstein, merely that the language-game of science is played? Even if it is impossible to do science without holding to a conception of objective truth, Rescher has still not provided an argument for doing science. He has simply given a 'transcendental argument' as to what has to be accepted before the practice of science is possible. We need the postulate of objectivity too before we can operate our conceptual scheme. As he says, 'The justification of this postulate lies in its utility' (1992, p. 265). Without this concept of reality, we cannot form any crucial notions which we need, such as those of truth, facts, and inquiry.

Rescher's pragmatism, however, means that he goes further than Wittgenstein. He does believe that science, by any test, provides the most reliable route to knowledge:

> The mechanisms of scientific reasoning clearly represent the most developed and sophisticated of our probative methods. No elaborate argumentation is necessary to establish the all-too-evident fact that science has come out on top. (1992, p. 178)

We have, it appears, to take the success of science on trust. Because it seems virtually self-evident, the only philosophical problem which can confront us, if Rescher is right, is how we can explain this success. It is like watching baseball and trying to deduce the rules and purposes of the game from the way it is played. We do not have to worry about whether it is a proper game, or worth playing. We are perhaps part of a large crowd in a stadium, and can take certain things on trust. Similarly in our life, we are present as members of a civilization built on the obvious success of science. We must merely work out what has to be the case to explain the situation. Yet the question remains whether we are attempting to explain the actual success of science, taking it for granted that it has succeeded, or to show what, if anything, is the grounding of science. Rescher is claiming that we cannot do science without thinking that we are making discoveries about an independent reality. Yet he has not shown that there *is* such a reality or that the practice of science might not be totally ill-conceived. His very use of the word 'tool' to refer to a concept which should supply the basis for the whole of science demonstrates this. The point should not be whether it is useful to believe in reality, but whether it is right.

Rescher's argument is in effect that it is expedient for us to believe in a metaphysical reality, since our science would collapse if we did not maintain this as a goal. He exemplifies the normal pragmatic preference for beginning with the agent's point of view. We each act in the world and have to draw out the presuppositions of our own practices. No other point of view could be accessible. Yet Rescher is in some respects an untypical pragmatist. By emphasizing the success of science, he is giving it a privileged position, and apparently begs the question as to why it should be counted successful. To say that it is evident that science has come out on top seems not much better than making a sociological observation about the state of our society. Why, though, should we rest content that scientific reasoning is accepted as the very paradigm of rationality? How can we be sure that science *is* successful? What counts as success? Rescher has made our practice primary, and by making objective reality a necessary presupposition of it he has deprived us of being able to use it as a source of justification.

Pragmatic Justification

There are many contemporary arguments, as we shall see, against thinking that we can have any independent purchase on reality, and some of them come from avowed pragmatists who are explicitly following Dewey. Not all of them, however, take the superiority of science as their starting-point. After all, if the existence of a practice is what is important, rather than its grounding in the way things are, we might also reasonably set out to validate other types of practice. Following the same method, we have only to investigate their necessary presuppositions if they exist as going concerns, and then posit the reality (or realities) they each assume as their starting-point. If a mind-independent reality is essential for the conduct of science, why does the same argument not work for religion? From a sociological point of view, religious practices are as central to much human activity as is Western science. On Rescher's argument, if religion needs a belief in a transcendental God as much as science needs the conception of an inquiry-independent reality, we appear to have proved the existence of God as effectively as we have upheld the notion of reality as a target for scientific investigation. Rescher himself sees this as a possible objection as follows:

> God is essential to the project of religion and worship: the external world is essential to the project of inquiry and cognition. But perhaps these entire projects are simply inappropriate. (1992, p. 265)

Certainly an atheist who wishes to uphold the status of science will be very dissatisfied with this. A form of argument has been introduced which could be applied to all practices, without differentiating between them. The floodgates will be open, if we are able to make metaphysical postulates to save any conceptual scheme that appears to be a going concern. No scheme can be ruled out on the grounds that it is based on ignorance, superstition, or plain honest error. Yet whatever the function of metaphysics, it certainly should not be the servant of existing social practices or even of epistemology. What we know and do must be grounded in what is the case and not the other way round. Starting from where we are may seem inevitable. It is indeed even tautologous, since where else could we start? As a methodology however, it hardly stops us from being irrational. The reason for this is simply that it gives a privileged position to our present beliefs. Yet they may need close examination, revision, and even rejection.

We can say that where *we* now stand provides the criterion for rationality. The inconvenient fact that other people stand in different places is beside the point. Our view must prevail. Others will, however, argue in a similar way, and it will be hard to avoid adding the caveat that at any rate it will prevail for us. This is a truism saying that we will think what we will think, but it is also hard to judge precisely who 'we' might be. Relativism has restricted truth either to individuals, in which case it is more properly termed subjectivism, or to groups, communities, societies or whatever. It is in fact always difficult to define precisely the composition of the relevant society, and the vagueness in terminology is not accidental. Matters are clear if we deal with geographically separated tribes or nations. In a modern society, however, there are so many cross-currents of agreement and disagreement that specifying who 'we' might be is difficult. If we look at the question of the respective validation of science and religion, many scientists in a laboratory on weekdays will no doubt attend church on Sunday. One cannot easily distinguish on a sociological basis the standards implicit in the practice of science from those implicit in that of religion. Whatever the theoretical or rational differences, there are not two separate groups of people, who can be easily identified. As a result, their practices may well tend to run into each other.

Rescher, himself, makes it clear that the comparison with religious practices is not a threat to science or to cognitive inquiry:

> The religious project is optional: one may simply decline to enter in. But the cognitive project is not so easily evaded. We must act to live, must eat this or that, move here or there, do something or other. (1992, p. 265)

It seems that our need to survive and to flourish in this world will immediately generate the need for rational inquiry. 'We cannot,' he says, 'act effectively without rationally warranted confidence in our (putative) knowledge' (p. 266). Our science is built on our communal search for such rational warrant, and that cannot take place without a commitment to the real world. Indeed such is Rescher's allegiance to science that he believes that it must eventually itself be able to explain how it is that creatures such as ourselves can develop a 'workable view of the world'. Thus he can say: 'The validation of scientific method must in the end become scientifically validated' (p. 266). Yet all this must be based on a metaphysical postulate which makes it possible.

Is religion optional in a way that science is not? Certainly in order to survive in the world, we need to make decisions about what it is like. People who do not see holes in front of them are liable to fall into them. So-called 'evolutionary epistemology' is built on this insight, as we shall see. Yet it is unclear that the whole edifice of Western science, including all the latest research into the nature of subatomic particles, can rest on this need for survival. It is all too obvious that much of science has itself resulted in even greater threats to human survival and to the survival of our planet as a viable ecosystem. The human need to live in the world was met over many millennia to a greater or less extent without any resort to the practice of science. Basic human cognitive abilities are one thing. Advanced science is most decidedly another. It could be argued that the latter is the natural outworking of a human rationality that has been developed to serve our basic needs. Why, though, should a set of practices which are very specific to a particular historical period, our own, be turned into the absolute standard of rationality? Why, too, should the near universal fact of human religion be relegated to the realm of the optional? It is hard not to conclude that Rescher, for one, has made up his mind about what is important in human life and decided what is 'optional' on that basis.

This is not to deny that there is some attraction in a pragmatic justification of religion, possibly with attendant transcendental arguments as to what must be the case to enable us to operate a conceptual scheme. The ultimate grounding of religion is as important as that of science. Yet one can certainly do without religion and many are beginning to question whether we may not all be better off doing without science. In the case of religion, there is the added complication that there are many different religions and religious practices. One of science's main claims to respect is its ability to command assent and agreement through the application of its methods in all cultures and parts of the globe. This is definitely not the case with religion. As a result, starting with human practices can quickly lead into the quagmire of relativism. William James asserted that 'On pragmatistic principles, if the hypothesis of God works satisfactorily in the widest sense of the word it is true' (1907, p. 299). James avoided relativism whilst recognizing the plurality of faiths by saying that we do not yet know certainly what type of religion is going to work out in the long run (p. 300).

However unsatisfactory this might be as a foundation for belief in God, it is an outlook which inevitably stems from James' insistence that the pragmatic method means 'the attitude of looking away from first things, principles, categories, supposed necessities and of looking towards last things, fruits, consequences, facts.' We should not therefore start with the concepts of reality or of God, but look to make sense of our experience, and old and new facts. This is intended as a method, and not a doctrine, and it is not surprising that self-proclaimed pragmatists disagree among themselves about, say, the differing merits of science and of religion. Nevertheless, just as pragmatism leaves religion with the unstable basis of human practices, it also destroys the possibility of a proper metaphysical grounding for any other human practice, including, in particular, that of science. The practice of science becomes, in a sense, its own justification, and this makes it look very unstable once faith in its efficiency and benefits begins to crumble. The removal of ontological foundations leaves science exposed to great danger when buffeted by adverse social forces. The same has already happened in the field of religion, where doubts about its metaphysical grounding make people reluctant to continue religious practices. Certainly our intellectual heritage, whether in religion, science, or any other field, cannot be rejected while sociological practices remain untouched. No one will continue to give their allegiance to science if a proper justification cannot be given for doing so.

Warranted Assertibility

John Dewey said that 'knowledge in its strictest and most honorific sense is identical with warranted assertion' (1938, p. 143). This notion of warranted assertibility has become increasingly popular, as a way of avoiding more metaphysical ideas of truth.

The idea is that the emphasis has to be placed squarely on the context of inquiry to see what warrant we have for what we say. As always with pragmatism, the process is one from the bottom up, from where we are, rather than from the top down, from some inaccessible reality. Rescher treads a similar path when he proposes that we do not evaluate an inquiry procedure by the truthfulness of its results, but rather approach the question of truth through the idea of warrant: 'One does not approach warrant by way of truthfulness, but truthfulness by way of warrant' (1992, p. 239). Warrant is the independent variable. As a result, a premium will be placed on the coherence of all our beliefs. Whatever the attractions of defining truth in terms of correspondence with reality, we are forced by such a definition to face the problem of the access we might have to reality. As the opponents of realism never tire of remarking, we cannot approach reality without having beliefs. There is no independent access, by means of which we can check our beliefs. The correspondence theory of truth, emphasizing that truth consists in the correspondence of beliefs with reality, is singularly unhelpful as a criterion of truth. It seems as if all we can do is achieve some kind of consistency between our various experiences and beliefs, and that is where the pragmatist starts. Yet the correspondence theory fares better when understood as a definition of what truth consists in. It then stresses the importance for truth of a proper relationship between true statements and beliefs on the one hand and the way things are on the other. This may not be a good epistemological guide, but in showing that truth is intimately connected with the idea of reality, it forces us back to the metaphysical problem of the nature of reality.

Any appeal to metaphysics, however tacit, would be enough to make pragmatists reject the idea of correspondence. The pragmatist ideal of piecemeal advancement is captured by an image borrowed from Otto Neurath, and is often quoted with approval by W. V. Quine. Neurath, one of the central figures of the Vienna Circle, pointed out that there was no fully secure starting point for science. He asserts: 'We are like sailors who have to rebuild their ship on the open sea, without ever being able to dismantle it in dry-dock and reconstruct it from the best components' (1983, p. 92). Significantly, he goes on to say that 'only metaphysics can disappear without trace'. On that, at least, positivists and pragmatists could agree.

The emphasis on warranted assertibility certainly changes the focus of our attention from the reality that is beyond our grasp to our actual practices of inquiry. Yet it raises the question as to what is meant by warrant, if we rule out any appeal to the way things are. For Dewey, what counted as knowledge was individually linked to particular, local circumstances. There was, he thought, no such thing as knowledge in the abstract. It was the product of inquiry, and inquiry is 'a mode of activity that is socially conditioned, and that has cultural consequences' (Dewey, 1938, p. 19). Understood in this way pragmatism will always turn away from the abstract and the general to concentrate on the concrete and local. Even the place of human reason itself is going to be under scrutiny. It certainly could not be a faculty which recognized logical, self-evident truth. As Dewey pointed out, we have realized that such 'truths' as those of Euclidean geometry are themselves derived from postulated axioms. The pragmatist will always be happier with an instrumentalist view of reason which works out the relationship between means and given ends. This picture will be of our finding ways to satisfy desires and needs. According to James, even the growth of knowledge is a matter of satisfying desire. He counts an opinion 'true' 'just in proportion as it gratifies the individual's desires to accommodate the novel in his experience to his beliefs in stock' (1907, p. 63).

Reason and reality have to be closely linked concepts. Dewey wished to repudiate any distinction between the subject and object of judgement. Since the rational subject was typically thought of as trying to obtain knowledge of objective reality, this meant a radical revision of traditional ideas. Without an objective reality to be known, there would be nothing to reason about, and without what has traditionally been termed the faculty of reason, we would have no way of discovering the truth. Amalgamating subjects and objects must in effect remove any possibility of reasoning about the world. It also stops us theorizing about the effects an independently existing world has on us. The world and our beliefs about it are collapsed into one another.

The pragmatist vision challenges the possibility of reasoning, at least as traditionally understood. Since science is itself the out-working of human reason, it also undermines science. Without the fundamental distinction between our beliefs and what they are about, science cannot proceed. Rescher saw this even from within the tradition of American pragmatism. Yet it is not enough to point this out, as one legitimate response to being shown the necessary conditions for the possibility of a practice is to give up the practice. The argument can destroy science as well as giving it the necessary underpinning. Indeed, attacks on the distinction between subject and object, between reason and reality, come from all sides, and sometimes they come from avowedly nihilist viewpoints. The tendencies of pragmatism certainly find resonances in a wide range of other philosophies. Nietzsche and the later Wittgenstein were equally suspicious of the independent power of reason. The result is that we are rooted in the practices of particular societies at particular times. Under the influence of Heidegger, Gadamer has emphasized how we are the creations of a particular historical reality. We are the products of a tradition and it follows that 'the idea of an absolute reason is impossible for historical humanity':

> Reason exists for us only in concrete, historical terms, i.e. it is not its own master, but remains constantly dependent on the given circumstances in which it operates. (1975, p. 245)

This type of view has clear affinities with Dewey's emphasis on the social setting of enquiry, although Dewey accepted Peirce's definition of truth as the product of ultimate agreement.

The explanation for Dewey's wish to attach reason to contingent local conditions lay largely with his revulsion for the Greek idea of reason being coupled with the perception of necessary truth, or the interaction of eternal essences. Too close a link between rationality and the understanding of *a priori* truth can be very debilitating. It will leave unanswered the question of how we make conclusions about the nature of contingent reality. Science is based on the insight that the world does not have to be as it is, and there has been a tendency to suppose that pure reason can deal with logic and mathematics, whilst rationality as such is unnecessary for making sense of our experience of the world. The ordering of experience is taken to be a degraded instrumental rationality. It is hardly surprising that with reason solidly on the side of deduction and logical inference, the idea that it can play an active role in the understanding of the world becomes highly problematic. As a result, induction seems to be cut off from any rational basis once the notion of any speculative reason which is not based on deductive logic has been ruled out. Reason and metaphysics need each other, and it is no accident that an attack on one is also an attack on the other. The more reason is rooted in a particular time and place, and is seen as the outcome of historically contingent circumstances, the less it can break free. Only when it is liberated can it conceive of what may be true of the world as such, rather than possessing a view of reality that has been historically created. When reality recedes as a 'thing-in-itself' and even eventually drops out of sight, a free-floating reason fails to latch on to anything that is invariant between different times and cultures.

Historicist views like those of Gadamer have been gaining currency. All understanding is related to a particular context. If, for example, our own understanding is thought not just to be influenced by the conditions of our own era, but to be totally formed by them, there is an insuperable problem as to how we could ever understand those who were moulded in other cultures and epochs. Without the ability to appeal to something that transcends time and place, we are all imprisoned by the assumptions and prejudices of one time and place. Reality has often been linked with the idea of invariance, of what stays the same across different viewpoints. Since the whole point of historicism is to link reality and viewpoint, what is seen from one viewpoint cannot then be identified with what is seen from another. What is judged according to one tradition, or in one society, could never be simply correlated with the judgements of another. There will be no common ground on which to stand, no raw experience on which we all depend, and no common human nature invariant across cultures (see Trigg, 1985, e.g. p. 102). The object about which we form beliefs is as much the creation of our present contingent circumstances as we the subjects are. Gadamer is typical of many when he claims that our very subjectivity and personality is formed by the society and language in which we are educated. He says: 'The self-awareness of the individual is only a flickering in the closed circuits of historical life' (1975, p. 241). Any idea of a metaphysical self that can rise above local circumstances is ruled out as firmly as the reality which the self might try rationally to understand. There have been those who imagined that science could somehow be exempted from these historicizing tendencies. Indeed, the apparent certainties arrived at by the empirical methods of science have seemed to stand in stark contrast to the shifting fashions of other forms of belief in society. This has proved, however, to be an unstable position to hold. Science has too obviously itself developed, changed and, dare one say it, succumbed to the fashion of the moment. As a result, it is all too possible to allege that it forms a tradition like any other, arriving in a particular place at a particular time, with no more ability than anything else to reach for a rational justification of its practices.

Ungrounded Practice

Pragmatism has no dealings with empirical sense-data, and is not inclined to give a privileged position to empirical 'facts' as against values. Dewey in particular was anxious to establish the world of value on a similar footing to that of other forms of knowledge. He did not wish to subordinate one type of practice to another. For him, ethics and aesthetics could be as objective as natural science. That might appear to be a gain, in that it allows human reason greater scope to deal with aspects of human life other than those experienced through the senses. The catch is, however, that so far from raising other practices to the status of science, the latter has itself been devalued. It is not very impressive to be told that ethics can claim as much truth as science, if the latter itself is no longer understood as the pursuit of knowledge for its own sake. As Dewey says, 'All reflective knowledge as such is instrumental' (1960, p. 218). We have to begin and end with the things of 'gross everyday experience'. Our reasoning must start with them and end by making a difference to them. It is not in any way an abstract search for truth. This emphasis on the practical consequences of inquiry resulted in an onslaught on traditional methods of education. Indeed any conclusions about large philosophical issues concerning reason and knowledge are bound to carry implications as to how children are taught. In particular, questions about the rational status of science are going to affect whether and how it is taught in schools.

It is no coincidence that John Dewey is renowned as one of the foremost influences on modern education. He was opposed to the inculcation of fixed conclusions, and used his philosophical position to attack the division between academic and technical education, based on the distinction between theory and practice. This illustrates the way in which philosophical conceptions of the nature of truth and reality, coupled with the scope and power of human reason, lie at the very heart of any idea of education. Dewey followed Plato in showing how ideas of knowledge must affect education. He was opposed to abstract notions of reason and knowledge, but it is significant that he was himself engaged in the rational pursuit of philosophy. Perhaps the difference such ideas can make to the lives of even the youngest school children is its own pragmatic justification. Yet the point is sufficient to indicate that philosophers who use their reason to undermine the power of reason are indulging in a high-risk strategy.

Of all subjects, philosophy has attempted to transcend the barriers of time and space. Philosophers in this century still argue about the theories of Plato and Aristotle as seriously as if they were put forward by contemporaries. The issues of truth and reality have seemed invariant, precisely because reality itself has been conceived as invariant across societies and epochs. Paradoxically the arguments of those Greeks, like Protagoras, who wished to deny this and espouse relativism are as intelligible to us as they were to their contemporaries. Once, though, forms of historicism become influential, they remove the basis for any idea of a reason that is not embedded in the concrete situation which produced it. Arguments for historicism, or for related views which root reason in particular practices, must be eventually self-defeating. They cannot themselves genuinely aspire to a truth that transcends all particular circumstances, or tell us what is true of all periods. Pragmatists may not necessarily acknowledge any affinities to historicism, although some assuredly do. They are, however, likely to face similar difficulties. Having ruled out metaphysical truth, arguments for pragmatism have an uneasy status. Unless human practices are judged unchangeable, it must be acknowledged that the very practices which produced pragmatical reflection may themselves change. In that case, arguments for pragmatism must be regarded as provisional, and revisable in the light of further practice. If pragmatism as a theory does not work when applied to current practices, then it will have to be modified.

Bertrand Russell complained that American commercialism obscured a love of truth, and he considered pragmatism to be the mere philosophical expression of this. Not surprisingly, Dewey could not accept this criticism and considered it of the same order of interpretation as saying that the French preferred dualism as a philosophical theory as 'an expression of an alleged Gallic disposition to keep a mistress in addition to a wife' (Schilpp, 1951, p. 527). Philosophical argument could easily degenerate into this kind of name-calling, but Russell had a point. Whether his diagnosis was accurate is irrelevant. Tracing a philosophical position back to its social roots is the inevitable outcome of pragmatism. In a sense Russell was playing his opponents' game. Any emphasis on the concrete and the particular will lead us to look closely at the practices and philosophy of particular societies. We can no longer talk of what is universally true. In fact, an instrumentalist view of reason must impoverish the whole idea of philosophy. If our ends are presented to us in our social practices, human rationality can do little to uphold or criticize them, but is their servant. We can only reason *from* the standpoint of a society. We can never reason *about* it.

This is all especially relevant when the status of modern science is questioned. Our society, however we wish to define it, is dominated in every way by the practices of the physical sciences. According to pragmatism, therefore, instead of being able to question the purpose of science, our reason has to be employed working out the implications of this commitment to science. Like empiricism, pragmatism seems especially suited to be the philosophy of a scientific society. The emphasis on what works seems suited to the apparent successes of science, even if it comes to be recognized that the idea of what 'works' is itself the product of particular social context. Yet empiricism provided a foundation for knowledge in the fact of human experience. The latter seemed to transcend particular social settings and could provide a rationale for our pursuit of science, even if not one rooted in metaphysics. Pragmatism's very emphasis on starting from where we are makes the idea of any justification for our more fundamental beliefs impossible. They are just the ones we happen to find most difficult to give up at a given time. In coherentist fashion we can judge one belief in the light of another, and hope to achieve consistency. The pragmatist would allege that this is all we can ever do. Our practical purpose provides us with our theoretical norms and that means that we cannot detach ourselves from our actual social setting.

Pragmatism can in fact go in two opposing directions. It can take stock of where we are and ask what must be true to explain the evident success of our practices. This is the path back to metaphysics. Rescher asks:

> What sort of evidence could *inquirers, constituted and positioned as we are,* possibly gather to *account* for the way in which inquiry procedures they employ actually work, and to legitimate their use? (1977, p. 1)

Rescher, himself is not afraid to talk of the uniformity of nature, at least as a regulative presupposition, and it is clear that even when we begin with our own practices, the question of legitimation can become pressing. Since, though, that can quickly involve a return to the very metaphysics which is anathema to pragmatists, a second route is perhaps the more typical path for contemporary pragmatists. It is to note that since we cannot escape from our historical situation, humanity should not be understood as one community. There is no such thing as a human rationality as such which can leap over the boundaries of time and space.

Pragmatists certainly could not conceive of a human reason which somehow echoes a deeper rationality that perhaps inheres in the very nature of things. Starting from where we are can also involve noticing that other people have started from different places.

Pragmatism does not have to be relativistic. It cannot easily ground itself in metaphysical truth but it does sometimes look forward to some final universal agreement. It can emphasize how our most important needs and desires spring from our biological nature, and not some historically constructed nature. It can in this way appeal to factors transcending individual societies. Yet at a time of rapid change, and in societies which cannot easily find an agreed framework of belief, the status of science is being more persistently questioned. No longer does it provide a standard other parts of society are expected to emulate. The temptation will then be for a pragmatist to stress the changing nature of our practices, and to resist the notion that somehow we are all being inexorably driven to some overarching final agreement. That in itself seems suspiciously metaphysical, and it is more rational for pragmatists to see the sheer contingency of agreements and social arrangements. Not only do they not have to be as they are, but in a few years time they might well have changed.

Without any metaphysical underpinning there is in fact nothing to agree about, nothing to constrain and guide us in our pursuit of truth. It is no wonder that practices are thought to shift arbitrarily. The very concept of truth becomes problematic. It used to be widely accepted that any belief or statement must be *either* true *or* false. This has become doubted, so that many would reject this so-called law of the excluded middle. Joseph Margolis sees in this rejection one of the main signs of relativism. He suggests that 'any doctrine counts as a form of relativism if it abandons the principle of excluded middle or bivalence . . . or restricts its use, so that, in particular sectors of inquiry, incongruent claims may be validated' (1991, p. 17). This is admittedly a weak form of relativism, but it does have large implications. It suggests that there is nothing against which different claims can be measured. In the case of science, it would suggest that our theories are under-determined by the world, or, worse, there is no sense in conceiving of one world independently of our theories. It suggests that there is no invariant order built into the structure of things. The world cannot be known by us, or, in Margolis' words, it is 'cognitively transparent'. This inevitably leads to the view that conceptions of real structures are at least as much the product of our society as of any world. This in turn should make cultural diversity unsurprising and leaves us with historical change as a brute fact.

It is a paradox that stressing the practical aspects of science, and the way we interact with things around us, can soon undermine the basis on which we deal with the world. The dismissal of metaphysics can at first appear liberating, but the future of science as a practice becomes problematic unless we believe there is something to discover. Ungrounded practices can become part of the flux of an unstable and essentially meaningless life. Margolis writes to espouse a form of relativism, but he does see the risks involved in the enterprise. He sums up a view which he considers we all instinctively cling to as follows:

> That man's nature remains unchanged through all the currents of changing history, and because of that, that there must be a final order hidden in change itself, an order adequate for saving the norms and the very meaning of human existence, for grounding our science, for giving our active interventions an intelligible and real space in which to move rationally. (1991, p. xii)

Relativism, as a doctrine, typically opposes any views of a fixed human nature, a rationality that can transcend the contingent circumstances of a particular society, and an objective reality. As such, it is opposed to any idea of fixed standards. Yet like all such views, it is notorious for being too ready to assume the truth of what it denies. Relativists are prone to accept the objective reality of constant change, and the impossibility of objective reality (see Trigg, 1973, for a sustained critique). They are liable to assert the truth of the assertion that there is no truth, and to point to the good reasons for everyone denying that reasons have force outside the society in which they gain currency. One of the problems about relativism is whether it is incoherent even to believe it, since belief surely suggests a belief that something is the case.

This is a recurring theme, but there are connections between attacks on the notion of a human nature, a rationality transcending local conditions, and an invariant objective order in physical reality. All these ideas presuppose something existing beyond the bounds of particular societies. Margolis claims that most philosophical theories in their various ways are now committed 'to the doctrine of the profound plasticity of the world' (1991, p. 120). The idea of there being any structures in the world or indeed any determinate world which can ground knowledge and rationality has been under attack from many disparate quarters.

The sheer 'contingency', not only of the world but of our ideas about it, has been stressed in a way that helps to explain great cultural diversity. The word 'contingent' is a slippery one, since it is often simply opposed to what is necessarily true. In this kind of context, however, it gains an even more radical meaning, whereby there is no reason why anyone should think one thing rather than another. It is in fact another instance of the tendency in philosophy to assume that the only alternative to logical truth is what is arbitrary, and happens by chance. The problem is always one of finding a role for rationality that makes it more than a haphazard product of a particular era, and less than the intuition of self-evident truth.

Pragmatism started with the insight that science works, and that it was unnecessary to produce any justification for that obvious fact. Yet even a pragmatic justification of the kind produced by Rescher seems barely sufficient. We should not merely appeal to what would have to be the case to make science work. We need to have some idea of what is the case. In order to underpin science we need to appeal to the actual structure of physical reality. It is at least conceivable that we are harbouring a major illusion about the success of science. Our ability to manipulate the world in certain respects could be a lucky accident, or the product of unusual local conditions in our part of the universe. It could be that in the end we can know no final assurance that this is not the case. This, however, is the path to total scepticism about the possibility of human knowledge. We can even be driven to wonder whether science actually works as well as we thought. Perhaps even that is an illusion. Pragmatism, however, refuses to take the threat of scepticism seriously, and thus fails to provide an antidote to it. We are still left wondering why science works.

3

The End of Reason

The Threat of Relativism

Pragmatists have often been accused of relativism. This is something they have in common with the post-modernist movement, which we shall shortly be examining. We have seen how pragmatists insist that we start from where we are, and must always insist that we use our local, parochial standards of rationality. As they will point out, there is nothing else we can do. We could not use the standards *others* consider appropriate unless we ourselves had a change of heart. All we can do is judge what seems appropriate from where *we* stand. Others in different positions will be doing the same from their point of view, so that there will be a divergence in judgements, and even different starting points. We will rate empirical evidence and scientific discovery highly, while other cultures, unless they have been overrun by Western science, will have different standards of what is to count as truth. We may not even count their procedures as rational, but that is because we have to be firmly based in one conceptual scheme. Others may have different concepts, but we *have* to judge as rational what *we* count as rational. The very idea of reason becomes firmly tied for us to the procedures of science. A form of conceptual relativism seems thereby to be adopted, but some pragmatists may deny that this amounts to full-blooded relativism. They may concede that concepts may vary, but hold it to be in the very nature of rationality, as we understand it, that we consider our position to be superior to others, if it is rationally established. A firm emphasis on our own standards implies that those who disagree are mistaken. The pragmatist would often agree that this is so but would add that of course this can only be how *we* view things, situated as *we* are.

The impossibility of detaching ourselves from our present perspective appears to make an alternative impossible. The idea of truth as seen without any particular perspective, a 'view from nowhere', is made to appear self-contradictory. We seem to need a perspective that raises us above all perspectives. Yet a strong form of relativism seems to beckon if we are imprisoned within one perspective whilst others adopt alien viewpoints. It does not seem possible that we should ever be able to grasp their concepts. Once we acknowledge the fundamental nature of a variation in perspective, so that even what counts as a rational standard might vary, the fact that we wish to affirm the rightness of our own standards will seem in need of further explanation. The paradox is that we should still wish to stress the correctness of our vision, while acknowledging that others are proceeding in analogous fashion from differing starting-points. Should we conclude that any criterion of truth is as good as any other? Does the acceptance of the fact of different conceptual schemes, with different criteria of rationality, weaken *our* hold on our ideas of what is reasonable?

It is possible to deny that others, sufficiently different from us, can have criteria of rationality at all, on the grounds that the concept of rationality is ours. This indeed is the first step to denying that there are alternative conceptual schemes. It may be alleged that either they can be translated into our terms, and are thus not genuinely different, or they cannot and will be so incomprehensible to us that we cannot even identify them as conceptual schemes in the first place. We may, however, allow some overlap, but deny that the standards of what is to count as rational are shared. The conclusion would be that they are not rational by our criteria. They would, though, still hold tenaciously to their own views. To be specific, we must face the question of what to do in a confrontation with those who do not accept that the physical sciences provide the standard for rational belief. They may, as Wittgenstein suggested, believe in oracles, or they may rely on mystical experience, revelation from a higher power or any number of other alleged sources of knowledge. The problem is whether this affects in any way our own allegiance to the standards of scientific investigation. Is it enough to insist that science is the source of rationality?

Science provides *us* with what we judge to be the appropriate criteria for judging truth. Those who repudiate science or the arbiter of rationality have a different perspective. We may deny that they are rational by our lights. Putting the problem in this way makes it look as if we are faced with a stark choice for or against science. This is the way the matter has often been portrayed. It is, however, one thing to assert that science is an example of human rationality, but it is another to identify the two. Even if science guides us in the discovery of the nature of physical reality, there can be other standards of what it is rational to believe that we can adopt. Science may lead to truth, but it does not have to be the only route. Many have become so enthusiastic about scientific method as a way of dispassionately winnowing out the distortions of personal prejudice, that they have identified truth with the operations of science. Yet this is to make the enormous metaphysical assumption that the reality to which science has access is the whole of reality. It is to assume that we, as humans, have no other source of knowledge nor any other means of reasoning. A scientific view of the world can narrow our vision unduly. We must not take it for granted that just because it is rational to practise science, it may not also be rational to pursue other forms of knowledge. Pragmatists can recognize this and acknowledge that moral judgements and practices can claim equal validity with scientific ones. They must, however, accept the practices as brute facts without any further grounding. They cannot give any justification for them beyond that they exist.

This question of justification lies at the root of dissatisfaction with the stress on arguing from where we are. The latter seems to be simply a dogmatism of the present moment. Whatever we happen to think at any given time is what we loudly affirm. There is a sense in which we can do no other. We can only think as we do think, and judge rational what we judge rational. Yet we cannot shirk the question whether we are right to do so. Being told that we can only think right what we do think right just returns us to an obvious truism. It does not address the deeper problem of whether we should rest content with where we are or should be willing to change our minds, if necessary. Should we be willing to examine and test our beliefs? The fact that they are ours for the time being is irrelevant. Too much emphasis on the fact of our acceptance of certain norms makes it difficult to explain *why* we accept them.

This is why pragmatism seems a close relation to relativism. One of the key points made by relativism is that whatever standards and norms are used, they can have no inherent justification. Their importance lies wholly in the fact that they are accepted by a particular group. Justification becomes transmuted into the fact of acceptance. What people believe is subordinated to the fact that they believe it. The very idea of justification being other than local and secondary to the adoption of a particular perspective is ruled out. We then cannot believe anything on the grounds that that is how things are. We cannot judge anything using rational criteria that are inherent in the very nature of things. Our standards just happen to have arisen at a particular time and place. They cannot reflect some wider reality. Pragmatists try to rescue themselves from the numbing winds of nihilism that can so easily start blowing. They reiterate the fact that if we think it rational to pursue scientific investigations, and consider our present standards to be the appropriate ones, that is the best we can do. We can do none other than continue with our present faith in science. Why, they would ask, should we be any the less willing to be rational merely because we recognize that rationality is a local, rather than a cosmic, matter? We can only do the best as we see it, and rationality constitutes by definition the best method we know for arriving at beliefs and judgements.

Yet the problem remains that we do not seem to have much of a justification if it can only be couched in terms we ourselves already accept. The whole point of rationality is that it is not a parochial concern, or an optional extra. It is not an institution of a particular society. Its nature has already been totally misrepresented, if it can be portrayed like this. The close identi-fication of rationality and science in some people's eyes has not helped. It would follow from this that as not all societies are based on Western science, not all are committed to rationality. Yet if rationality is linked to the ideas of truth and reality, science is only rational because it provides a reliable method for gaining knowledge. Science does not constitute rationality, but consti-tutes one human response to the question how we can gain knowledge and understanding. We may be scientists *because* we are rational but shall not be counted rational merely because we are scientists. If this is right, a non-scientific society could claim to be rational if it looked for grounding for its beliefs, and was concerned about their truth.

Rational justification must of its very nature transcend the presuppositions and prejudices of any particular community. If we are justified in holding certain beliefs or adopting a particular method for testing them, such as that of empirical science, that justification should hold good for others in other societies whether they realize it or not. Genuine justification must be rooted in the character of the world in which we are placed, and not in the character of the people to whom the justification appears valid. In the latter case, there would be no distinction between apparent and genuine justification. We may think we are justified in our beliefs when we are not, but the realization that this is so should encourage a wholesome humility and a willingness to examine alternative positions. There is an unaccountable belief that rela-tivism must somehow encourage tolerance, while realist positions are essentially intolerant because they insist that objective truth is at stake. Those who talk of truth in the latter way are often portrayed as crypto-Leninists or supporters of the methods of the Spanish Inquisition. Certainly horrors have been perpetrated in the name of truth. It does not follow, however, that if I believe I am right I will want to impose my views on those who disagree, particularly if I believe it to be objectively wrong to do so. In fact the knowledge, and warning, that one may be mistaken is intrinsic to any realistic outlook. The idea that we may be able to learn from others because they know better than us only makes sense, given the realist distinction between what we think and what is actually the case. Relativists, on the other hand, cannot allow that there is any rational way of resolving basic divergences in belief. The appeal to reason is one they cannot make. They may not care what others believe, because they believe there is no objective truth which could fail to coincide with their own position. They may, however, care passionately, and arbitrarily impose their views by force in an exercise of brute power. There is no reason to stop them which could transcend their own particular prejudices. Relativists could then as easily be arrogant militarists as tolerant liberals. Liberalism itself can only be safe if it is secured by rational principles which can override the idiosyncrasies of particular societies.

The most corrosive aspect of relativism and of the pragmatist insistence on the validity of our present position, is that both deny that justification can extend beyond present prejudices. It arises from them and serves to reinforce them. There is no way in which we can wonder whether we may be wrong in being where we are, even if that is where we have to start from. Indeed pragmatists count it a virtue that they excluded all possibility of sceptical doubt. When, however, relativism, and pragmatism, dissolve the distinction between what is reasonable and what we judge reasonable, it is a sign that something profound is happening to our understanding of our place in the world. We are no longer constrained by the nature of reality. Our reason is no longer subjected to standards that inhere in anything beyond ourselves. Indeed reason has no force beyond what we give it. This may seem inevitable. I will not act for a reason I do not acknowledge, even if my acceptance is subconscious. I will not rest my beliefs on evidence I do not accept. Yet the idea that there are rational standards which we should accept, if only they were presented to us, is important.

Rationality does not just consist in what we now happen to believe, but points to a standard which ideally ought to be accepted by everyone, because it is the way in which we all become accommodated to reality. People, in whatever society, who are not rational, fail to judge things correctly, unless they are very lucky. It is hardly surprising that the ideal of rationality is jettisoned, if the notion of reality is abandoned. Yet the whole point of even holding to our present beliefs is put into question, if we no longer have to worry about how far our beliefs are rational, and whether we are living in accordance with the way things are. The pragmatic insistence on the appropriateness of present standards rings hollow. When we cannot give a justification of our position to those who diverge from it, we have to face the question whether we can justify it to ourselves. Of course, when others do not agree with us, they may not in fact accept any purported justification of our views. The point is, however, whether we can appeal to some feature of reality which they too ought to recognise. If we all live in the same world, reacting to the same things, are guided by reason, and have proper information, we all ultimately ought to agree. We may not, particularly in complicated matters when we have imperfect knowledge, but even so, rationality ought to play its part in drawing us gradually together. This is the kernel of truth in Peirce's emphasis on the importance of final agreement between scientists. The agreement itself is not the most significant point. The fact is that if they all gain complete knowledge of the way things are, they are bound to agree.

Whatever our beliefs, they should be anchored in reality. We can only claim justification with an appeal to something independent of our beliefs. If physical science is a source of knowledge, it must give us insights into the workings of our physical environment. In such circumstances, those who reject science will thereby demonstrate a lack of interest in the world they live in and have to deal with. The relativist and pragmatist refusal to allow the idea of an independent reality fails to give us any motivation for changing our current beliefs, or even for keeping them. The fact that they are 'our' beliefs is not enough. Once, though, the fact of belief is thought more important than its content, the fact that we believe something appears to be its own justification. We can no longer ask why beliefs are accepted, let alone whether they should be accepted by anyone else.

Post-modernism

American pragmatism has been a powerful influence in philosophy, and by attacking metaphysics, it has helped to remove the props on which our knowledge rests. It is, however, only one strand among several which appear to put our practices in a privileged position, while denying that we can give them any rational justification. The later Wittgenstein is another source, and Jean-François Lyotard, one of the proponents of 'post-modernism' uses the Wittgensteinian image of a game, which we have already encountered, when he deals with the question of legitimation in modern science. He points out that science leaves behind 'the metaphysical search for a first proof or transcendental authority'. He goes on to say:

> It is recognised that the conditions of truth, in other words the rules of the game of science, are immanent in that game, that they can only be established within the bounds of a debate that is already scientific in nature, and that there is no other proof that the rules are good than the consensus extended to them by the experts. (1985, p. 29)

We are left with no way of answering the question why the game should be played in the first place. As Lyotard points out, another consequence is that science cannot set itself above the language-games, whether of aesthetics or anything else. He says that 'the game of science is thus put on a par with others' (p. 40). Indeed he says that it is 'an important current of post-modernity' that science plays its own game but is incapable of legitimating others. According to post-modernism, there can be no overarching reason, no one true way of looking at things. The search for justification or legitimation was, for Lyotard, a particular feature of the *modern* era. Indeed he says: 'I will use the term to designate any science that legitimates itself with reference to a metadiscourse . . . making an explicit appeal to some grand narrative' (p. xxiii). The impossibility of stepping outside all particular language-games or frameworks is a major part of the post-modernist vision. There is no such thing, it is argued, as a language above all languages, or some story above all stories or 'grand narratives' that would provide the rational justification for the knowledge which we claim. Post-modernism is, in fact, defined by Lyotard as 'incredulity toward meta-narratives'

(p. xxiv). He continues: 'To the obsolescence of the meta-narrative apparatus of legitimation corresponds, most notably, the crisis of metaphysical philosophy and of the university institution which in the past relied on it'.

Like post-modernism, modernity itself is a somewhat hazy notion. The advocacy of rationality is certainly dismissed by post-modernists as a crucial aspect of modernity. The modern outlook tended to associate science with reason, and it is not surprising that suspicion of the former has led to attacks on the latter. The view of reason as encouraging unanimity between minds which are confronted with truth is itself described by Lyotard as 'the Enlightenment narrative' (1985, p. xxiii). It is thus dismissed as a product of a particular historical period, and a feature of the modern age which has now been superseded. On this understanding post-modernism is an attack on notions of reason and truth. It dethrones science by attacking the very human rationality which has produced science. Some may welcome the conclusion that science cannot monopolize human reason, but is only one language-game amongst many. They might, however, be more reluctant to accept that science has only lost its authority because all human reason has been rendered impotent. We can no longer legitimate any of our activities.

Post-modernism joins with other influential contemporary views in stressing that there can be no privileged position. We are all situated in contexts which cannot be transcended. There is no way of grounding our beliefs and no one method for obtaining knowledge. As a result the possibility of any knowledge is challenged. At first sight, resistance to domination by science can appear as liberating as does opposition to the stultifying conformity imposed by the slab-like structures of the modernist movement in architecture. Soulless tower blocks of flats and bleak office skyscrapers have seemed to be the expression of a culture which believed in the domination of science. It looked as if human nature could be moulded by the imperatives of science and technology expressed in concrete, rather than architecture serving the unchanging needs of people. Post-modernist architecture, prevalent from the 1980s on, has tended to build on a more human scale and has not been afraid to borrow styles from different historical periods. It revelled in diversity more than coherence. Just as its philosophical counterpart refused to accept the dictates of scientific method as the criteria of truth, so architecture refused the straitjacket of any one particular style, particularly that demanded by the technology of steel and concrete. All the styles of history were judged suitable for imitation and indeed some have appeared in playful juxtaposition on the same building. Classical pillars could be transmuted into something quite different as they progressed along a façade. Post-modernists were happy to provide useless ornamentation, in contrast to the stern demands of function emphasized by the modern movement. This functionalism meant that a façade which did not reflect the inner structure of a building was judged positively immoral.

The triumph of science is no longer celebrated quite so stridently in architecture. The element of frivolity and playfulness in post-modernist designs may have greater appeal than the grandiose concrete buildings of modernism. When, however, scientific claims to a monopoly of truth are overthrown, playfulness and irony, leading perhaps to cynicism and despair may be all that is left. It is perhaps significant that even in architecture, appeals are often made to certain styles, particularly classical ones, as having a value which transcends particular cultures, and which are particularly suited to meeting basic human needs. This presupposes something which post-modernism would deny, that human nature remains constant across time and culture. (see Trigg, 1982) Are human needs more than the creation of particular societies? This takes us back to the controversies about realism, since the conception of an essential human nature, detachable from any social context, is itself realist. It suggests that our real nature may even be different from the way in which our own society sees it. We all may, on this view, be committing major errors in failing to see, let alone meet, perennial human needs. This is no criticism, however, that can be made of modernism by post-modernism, since the latter denies realism and doctrines about the essential nature of anything. It glorifies diversity, and stresses the surface of things rather than the importance or relevance of any hidden depths. If no style is better than

any other, none can be worse either. There is then no basis
for condemning the concrete deserts that have laid waste our
modern cities. Post-modernism, in architecture at least, signals
a change in fashion, rather than a principled stand against 'in-
human' design.

As a cultural movement, post-modernism has had a wide influence in the arts as well as philosophy. It has been particularly
lively in the field of literary theory, in connection with the reading of texts. The emphasis there has been on the preconception
of the reader which governs the reading of the text, rather than
on any idea of the text having an intrinsic meaning, even that
given it by its author. The emphasis on the vagaries of interpretation inevitably undermine any idea of truth. The pursuit of
reason is reduced to the exercise of prejudice. In fact no text can
be regarded as intrinsically superior to any other. It is not surprising that the idea of a canon of 'great literature' is under
attack. According to post-modernism even the works of Shakespeare cannot be judged to have any intrinsic merit. All is interpretation. In the same way the position of science as a purveyor
of truth is suspect. There is a loss of faith in it both as a source
of legitimation, and as an intellectual force which has itself
received legitimation.

The loss of the idea of science as the sole guarantor of truth
has its counterpart in the loss of confidence in the modern style
of architecture which was itself inspired by science and technology. It is mirrored in the rejection of traditional standards of
literary criticism. Pluralism, understood as the availability of
alternative standards, each equally acceptable, seems to beckon
in every sphere. Once science, though, is no longer thought to
be the acme of human reason, and indeed is considered to have
no part in rationality at all, anarchy beckons. Science has been
too arrogant in the past, but just because its pretensions to a
monopoly of truth are queried, there is no need to undermine
science as a whole by destroying the very ideas of reason and
truth. The anarchic jumble of styles which typifies post-
modernist architecture at its worst could become a potent sym-
bol of the total confusion into which our society has fallen. The
philosophical debates about realism, relativism and the status of

science have a fundamental significance for the whole of society. Questions about truth, however wrapped up in philosophical argument, will always have a relevance for wider cultural issues. As we have seen, the way such issues are resolved eventually determines the shape of the buildings constructed in our cities.

The rejection of science's claims to truth stems from the rejection of truth as being anything other than what is accepted by particular groups of people at particular times. Yet the distinction between apparent and real truth and the desire to conceive of truth as a way that transcends historical contingencies is deep-rooted. Heidegger blamed this impulse to believe in some form of absolute truth as belonging to 'those residues of Christian theology within philosophical problematics which have not as yet been radically extruded (1962, p. 272). Talk of 'eternal' or 'absolute' truth encourages such a view, but the truth at stake is much simpler. It is objective truth, namely what is the case whatever individuals or cultures happen to believe. The assumption is that it is not the fact of belief but what the belief is about that makes something true. The truth of physics cannot be constituted by the thought processes or concepts of a particular historical society. If physics is to claim truth, it must not be on the basis of who accepts it. The question of legitimation must remain.

Solidarity and Reason

Heidegger's reference to theology in connection with truth was significant. It has indeed sometimes been remarked that atheists have transferred allegiance from one religion to another if they put their faith in science as an instrument of progress. There are overtones of a belief in salvation in such a commitment, as well as an allegiance to truth. As Richard Rorty has pointed out, 'a reaction against scientism led to attacks on natural science as a sort of false god' (1991a, p. 33). His response is that there is nothing wrong with science. He continues: 'There is only something wrong with the attempt to divinize it, the attempt characteristic of realistic philosophy.' Yet the identification of realism with scientism is questionable. Anyone who believes that science is the sole source of truth must be a realist, ruling out alternative viewpoints as false. Science, it is thereby claimed, tells us how things are, while other beliefs such as those of religion do not. A conception of reality as independent of beliefs is being invoked to give the scientistic thesis its bite. Although scientism defines reality in terms of the discoveries made in accordance with scientific method, realism and scientism are independent. The first move is to say that reality is independent of our conceptions, and the second is to identify it with the outcome of scientific investigation. One could accept the first and not the second.

There is a certain instability about linking reality too closely with actual science. It demonstrates a rational procedure for gaining knowledge, and reality is brought within our epistemological grasp. Yet doing so appears to involve defining reality in terms of actual or potential human capabilities. 'Science' is not an abstract metaphysical term in the way that 'reality' perhaps is. It refers instead to practices which are historically specific. It concerns what we, or people like us, can achieve situated in the kind of society we live in. Science is a human practice related to particular methods engaged in with particular instruments in Western society. Rorty would accept this and use it as ammunition with which to shoot down forms of scientific realism. Attacks, however, on particular conceptions of epistemology are not attacks on metaphysical views. The whole point of the latter is that they purport to concern what lies beyond the relativities of human existence.

Rorty himself would assert that metaphysics, as well as epistemology, must be understood as the expression of particular social practices. Like so many he is eager to place every emphasis on the fact of acceptance of certain beliefs and to ignore what they are about. In this, the influence on him of American pragmatism, the later Wittgenstein and contemporary trends in European philosophy stemming from Heidegger's historicism, all combine. Rorty argues: 'All the traditional metaphysical distinctions can be given a respectable ironist sense by sociologizing them – treating them as distinctions between contingently existing sets of practices, or strategies employed within such practices, rather than between natural kinds' (1989a, p. 83).

Rorty is opposed to the idea that there are basic distinctions between things built into the very nature of reality. The idea that there is an inherent structure waiting to be discovered is anathema to him. We cannot even consider ourselves 'human' in any substantial sense, since human nature is precisely one of the natural kinds that he rejects. Indeed he also rejects the idea of science as a natural kind. What he means by this is that scientific explanation is explanation of a particular sort, which is somehow related to reality in a way that other forms of enquiry are not. The ambition to find a clear way of demarcating science is linked, he considers, with a desire to find an account of the way in which human faculties are linked to the rest of the world. This would be a metaphysical account, or, as Rorty says, 'better yet', a physicalist one (1991a, p. 61). He enlarges on this by saying that this would be an account 'in which "reason" is the name of the crucial link between humanity and the nonhuman, our access to an "absolute conception of reality", the means by which the world "guides" us to a correct description of itself'.

The alleged links between science, reason and reality are broken by Rorty. Science cannot be given a justification in the metaphysical sense once its privileged status as a way of systematizing reality is denied. Science cannot set itself above other human practices once it is accepted that it does not have a unique claim to be the out-working of human reason. Yet once again, an attack on the pretensions of science is run together with a global attack on the possibility of rationality. If science is thought to *be* human reason *par excellence*, it follows that when reason is shown to be rooted in the contingencies of its context, so must science, and vice versa. Scientism certainly cannot be embraced without a strong conception of human rationality. Yet to put one's faith in science in this manner is to make a large investment in the local success of a particular manner of investigating the world. It is dangerous to restrict the practice of reason in this way.

Rorty refers to Dewey's pragmatist distaste for metaphysical notions (1991b, p. 20). He gives as examples 'the supra-sensory world, the Ideas, God, the moral law, the authority of reason', which he considers are all 'dead metaphors which pragmatists can no longer find use for'. In contrast he believes that notions such as progress, the happiness of the greatest number, culture and civilization do still have a point in a democratic community. However, as Rorty admits, Dewey can provide no argument against the one list or for the other. The explanation is, Rorty believes, that Dewey is not scientific enough to think that there is some neutral philosophical standpoint which would supply promises for such an argument. This remark clearly demonstrates Rorty's equation of science and reason, since the quest for a neutral standpoint, however forlorn, certainly need have nothing to do with science. Yet for Rorty only science could aspire to the kind of objectivity which rationality appears to demand. Neutrality and scientific objectivity are the same for him. The demolition of the one entails the removal of the other.

Rorty's invocation of the democratic community is significant. Elsewhere he refers to the question 'whether there is anything for philosophers to appeal to save the way *we* live now, what we *do* now, how *we* talk now – anything beyond *our* own little moment of world-history' (1991a, p. 158). Much depends who 'we' are understood as being, but once the distinction between subject and object is obliterated, we can no longer be understood as centres of rationality each striving to obtain knowledge of reality. Instead we are the product of historical contingencies, and rationality cannot aspire to involve any ahistorical principles. Objectivity can no longer be associated with truth. Instead, Rorty claims, 'the desire for "objectivity" boils down to a desire to acquire beliefs which will eventually receive unforced agreement in the course of a free and open encounter with people holding other beliefs' (p. 41). The familiar pragmatist emphasis on starting where we are has to restrict our vision, so that agreement rather than truth becomes our aim. We may be able to widen 'our' number by bringing others in through agreement. Rorty is emphatic, though, that we cannot rise above all actual and possible communities. He says: 'We cannot find a skyhook which lifts us out of mere coherence – mere agreement – to something like "correspondence with reality as it is in itself". Everything we consider real is, after all, what *we* consider real' (p. 38).

Science appeared to provide a way out of this morass, but Rorty argues that science is not a model of human rationality at all, but of 'human solidarity' (1991a, p. 39). He suggests that we must consider ourselves above all members of a community, instead of rational thinkers trying to gain contact with a reality that is not the mere product of negotiated agreement. Civic virtues should then replace the alleged intellectual ones. A community even of scientists does not exist as a means to some further end such as the pursuit of truth. The formation and maintenance of such a community should be goal enough. Unless, however, the community of scientists is considered of special importance because of its dedication to truth, or at least truth in a certain area, there will be the problem why anyone should want to join or foster such a community. Why, for instance, should the State, as the wider community, give scientists research grants?

Rorty is no doubt right to think that scientists should 'no longer think of themselves as a quasi-priestly order' (1991a, p. 44). It is, however, one thing to challenge science's monopolistic claims to truth, and another to say that scientists are just members of a community, and, moreover, one which can claim no philosophical foundations. Yet the inevitable results of the removal of the distinction between subject and object is that reason and truth are also put in question. There is no longer room for a distinction between what a person believes and what is actually true. Rorty is adamant that the idea of enquiry converging on the truth 'out there' should be abandoned.

Like Heidegger, Rorty hears echoes of religion resonating in any insistence that there is a single point on which all beliefs will converge. He says of this image that 'it seems to us pragmatists an unfortunate attempt to carry a religious view of the world over into an increasingly secular culture' (1991a, p. 39). There will be an ever expanding repertoire of alternative descriptions, but not, as he puts it 'The One Right Description'. He says: 'Such a shift in aim is possible only to the extent that both the world and the self have been de-divinized' (1989a, p. 40). Neither speaks to us in its own language, nor has its own intrinsic structure existing independently of the way it is described. Such a view makes it entirely unclear what is the purpose of science. The elaboration of multiple vocabularies, which do not purport to be about anything, and are not constrained by reality, seems a pointless activity. Part of the trouble is that like many post-modernists, Rorty is unable to conceive of any reality apart from language. He can only conceive of the self and the world in linguistic terms. There is nothing beyond the language or behind the text, underpinning, explaining and justifying. As a result, when there are competing descriptions, or alternative languages, there is nothing to which we can appeal against which they can be measured or judged. All we can hope to do is promote a civilized conversation between the holders of the different vocabularies. That, however, presupposes they will be able to understand each other and compare vocabularies.

The view that objective truth is part of a religious outlook is only half right. Religion needs truth but does not have an exclusive claim on it. Any genuine belief in God must root itself in a vision of the way things are. The argument between theist and atheist is *about* something, about the nature of the world. God, if He exists, is not a physical object, but must be objective, and not the creation of human minds or language. The anti-realist assertion that God does not have an independent existence, which gives meaning to the whole of reality, is itself an atheistic denial of God. There are those who wish to give a re-interpretation of religious belief in the light of post-modernism and its emphasis on language. Don Cupitt remarks, for example, how from Hegel onwards 'human reason was increasingly brought down into history and language' (1991, p. xii). This relativizing of reason has enormous repercussions for religious belief, since it removes the possibility of conceiving of reality beyond language. Since God has been by definition the Supreme Reality, and ineffable, beyond the full grasp of humans, this destroys any traditional faith in God. Indeed, Cupitt says that 'at its core, belief in transcendence or in God was belief in the possibility of mastering language from a standpoint outside it' (p. 142). Like Rorty's 'skyhook' to reality, this is made to appear an impossible scenario. Cupitt's position, however, merely serves to reinforce the fact both that religion has traditionally depended on a notion of objective truth, and also that its opponents need the same concept. Radical re-interpretations of theism completely undermine traditional understandings of God. In rejecting the admissibility of such claims to truth, writers such as Cupitt can only be understood as claiming the truth of a position which denies the possibility of any transcendent reality. The dispute between realist and anti-realist is in fact itself a dispute about the character of reality. It is an argument about what is true and false. A world in which there is no objective truth, but only 'stories' is one in which nothing is false or illegitimate. The language of the realist is as good a way of speaking as that of the anti-realist. Cupitt's own opposition to the realist understanding of God betrays a passionate commitment to truth.

Reason and Culture

The idea that our beliefs – whether religious, scientific or whatever – have to be measured against something beyond them is essential if they are to have any purpose. Pragmatists are, however, uneasy about being beholden to anything beyond human influence. Rorty says that 'pragmatists would like to drop the idea that human beings are responsible to a nonhuman power' (1991a, p. 39). Yet this power is not identified as God, but 'as a reality which is more than some community with which we identify ourselves'. What is being challenged is not so much theology as the possibility of philosophy, as it has been traditionally understood. Indeed, since philosophy as well as science, involves the exercise of human reason, attacks on the possibility of reason, or on the meaningfulness of truth apart from what happens to be accepted, are inevitably attacks on the scope of philosophy. The very idea of giving a philosophical justification for the practices of a community will he laughed at. As Rorty claims, 'edifying philosophy aims at making a conversation rather than at discovering truth' (1989b, p. 373). Philosophy becomes politics. The art of achieving consensus will be the nearest we will be able to get to the discovery of the nature of reality.

If philosophy is left with any task, therefore, it is that of prolonging a conversation rather than providing a justification. When communities cannot be grounded in anything beyond themselves, our sole task will be getting along with each other as best we can. Agreement is the most we can hope for, coupled with tolerance of those who disagree with us. Rorty makes no secret of the fact that he writes as an American liberal, who believes in persuasion rather than force. His ideal society has 'no purpose except freedom' (1989a, p. 60). Nothing, however, could show up the poverty of such a view more than its inability to give a justification of the society which produced it. Whatever the attractions and strengths of liberal democracy, it seems highly unsatisfactory if it cannot be defended rationally. For those in the world who are not American liberals, Rorty's emphasis on the contingency of all social arrangements and practices is hardly an advertisement for his own beliefs. His only defence of democracy could be that it is what he is used to. Any argument he might produce for its superiority would have to be grounded in something external to the particular arrangements of a particular community. An appeal to human nature might be relevant but Rorty can have none of that. The more he stresses the primacy of community, the more it appears that any rational examination of the nature and function of community is going to be impossible. Yet all the time he appears to be offering arguments as to why human reason is impotent. His attack on reason looks self-defeating. His 'edifying' philosophy seems to aspire to rational argument, even if its only conclusion is that rational argument is impossible.

Pragmatism's stress on re-building the boat which we are in looks more plausible as a description of the human predicament than as a description of the situation of the members of particular human practices. We cannot be other than human, but whatever particular practice we are involved in may always be one of several options. Liberal democracy is far from being the only possible, or even actual, political arrangement. The key question must be why someone far from the United States of America should be persuaded that democracy is a preferable political arrangement. It is perhaps understandable that someone from the United States should be schooled in the idea that truth should not be publicly established but is a matter of persuasion in the competition of ideas. In such a variegated society, with, for example, its tradition of religious freedom, there may be sound reasons for such a position. To transfer it from the particular, historically-produced circumstances of one nation and to make it into a philosophical principle of general validity seems dubious. This is all the more so in the hands of Rorty, who is opposed to the idea of philosophical principles, let alone ones with a force that can transcend local circumstances. His philosophical views reflect the society of which he is a member. For them to have relevance for the citizens of other nations, he has to present rational argument which has a more than local force.

Rorty, however, is not only a pragmatist but an avowed post-modernist, who agrees with Lyotard in distrusting 'meta-narratives', 'which purport to justify loyalty to, or breaks with, certain contemporary communities' (1991a, p. 199). It seems as if relativism is as inescapable for Rorty as for all post-modernists. Like many philosophers, however, Rorty acknowledges that relativism is self-refuting, asserting as true that there is no truth. He is therefore sure that he himself cannot be a relativist and argues: 'There is a difference between saying that every community is as good as every other and saying that we have to work out from the networks we are, from the communities with which we presently identify. Post-modernism is no more relativistic than Hilary Putman's suggestion that we stop trying for a "God's Eye View".' (p. 202).

Certainly we can consistently claim, from whatever position we are in, that our view is best and try to persuade others of this. The trouble is that reiterating the worth of American experience is not enough to engage the reason of those in other cultures. We may understand why, given his background, Rorty says what he does. The question is whether it is relevant to those in other societies. He can, *ex hypothesi*, make no appeal to anything transcending the circumstances in which he happens to have been placed. He cannot appeal to rational principles, to the nature of reality, or to the demands of human nature. No one imagines that we can abstract ourselves totally from all human traditions and judge them in the light of pure reason, like an omniscient God. We are deeply influenced by particular histories and traditions. The key issue, however, is whether this tells us the whole truth about ourselves. Many would agree that our common humanity is also an important factor. In the present context, however, the problem is whether we can abstract ourselves even from the very situation which has helped to mould us. Can we become sufficiently detached to reason *about* our tradition as well as simply *from* it? For this to be possible we have to be more than the product of the various biological and cultural influences that certainly help to mould us.

The concept of rationality suggests that we can question our heritage even if we may not be able to escape it completely. Yet standing back from a culture and being willing to repudiate it must be tantamount to denying ourselves if our culture gives us our identity. Reason is irrelevant if a culture is not grounded in anything beyond itself. Indeed there is then no such thing as rationality. Instead we are merely the sum total of our beliefs and desires, unable to distance ourselves from them, in order to justify or criticize them in terms which could claim validity beyond the confines of a particular community. Rorty believes in fact that there is no 'centre' to the self, but that in pragmatist fashion, 'there are only different ways of weaving new candidates for belief and desire into antecedently existing webs of belief and desire' (1989a, p. 83). He wants to maintain the distinction between using force and using persuasion, but his attack on traditional areas of reason makes it unclear why force should be ruled out. Presumably he would say that 'we' Americans happen not to hold with such conduct. As a political principle, that is less than persuasive.

Rorty says that 'metaphysics – in the sense of a search for theories which will get at real essence – tries to make sense of the claim that human beings are something more than centreless webs of beliefs and desires' (1989a, p. 88). In the distinction between subject and object, the role of the reasoning subject has to be as crucial as that of the object it is trying to understand. There is little point in establishing the existence of an objective reality, independent of our understanding, if we have no power to reason about it. Without reality there may well be no genuine truth, but unless we are rational subjects we will not gain knowledge except through a lucky accident. We could happen to be caused to believe what is true, but reasoning deliberately on the basis of evidence towards justified belief requires those who are able to transcend their own particular circumstances, and not be bound by any particular causal history.

Rorty takes 'conversation' as the ultimate context within which knowledge is to be understood. All we can do is to note the changing standards of what is to count as knowledge. Philosophy becomes merely the history of ideas. Yet the paradox is that even then we will be distancing ourselves from what we are describing. Historical reality is treated as an object with which we have to come to terms. The very project of cultural history is itself parasitic on the dichotomy between subject and object. Rorty's 'conversation' will itself necessarily be an exercise in coercion. There are different ways of achieving agreement and even of persuading others. Not all involve the free and glad acceptance by others in the course of rational debate of what they have come to see as true. Reason must be able to transcend its immediate context. Otherwise it is true that the forging of consensus will be all we can achieve. Politics will replace metaphysics. The prime question will then inevitably be how disparate groups can live together, and not how knowledge can be obtained.

4
Science and Naturalism

Philosophy and Science

We may hope for somewhere neutral to stand in order that we can reason about our beliefs. When this seems an impossible demand, the possibility of unprejudiced reason becomes suspect. The pragmatist impatience with the hunger for justification reflects something of this. We have to use our present beliefs as our springboard, instead of somehow standing apart from them, so that we can deliver judgement on them. It is often claimed that doubt should stand in as much need of justification as acceptance. Otherwise the difficulty of justification will inevitably mean that scepticism will win. For some, the apparent impossibility of a 'God's-eye' view, from which we could see things as they are, means the death of metaphysics. We certainly need an assurance that there is a reality in which our beliefs, if true, can be grounded. Yet, as will be argued in chapter 5, it does not follow that we need to have a complete knowledge of that reality, or even think that such a knowledge is attainable. This slide from reality to an omniscient view of it seems to involve a switch from what there is to how it can be verified, from metaphysics to epistemology. The hankering after some viewpoint which grants us total knowledge is in essence the desire for complete verification. The ghost of the verification theory still seems to haunt these controversies.

Philosophy does not need the possibility of omniscience or an assurance of verification to talk of the need for grounding beliefs in reality. Nevertheless those of a pragmatist disposition who wish to begin with what is currently accepted as true will undoubtedly look to modern science. They will hold that philosophy should give up all thought of standing apart from science in order to justify it. Instead it should accept the current findings of science, so that philosophy and indeed human reason, as such, should be subordinated to contemporary science whatever it may happen to be saying. Philosophy is the handmaiden of science rather than its rational underpinning. This is because science is regarded as the best account of the world we have. Indeed, philosophy, as the exercise of human rationality, is itself, it can be argued, open to scientific investigation. Yet since the argument that this is so is itself intended as a rational claim, we will again be faced with the problems of reflexivity.

If science is to be regarded as the ultimate arbiter of truth, much hangs on what is meant by 'science' and indeed which part of science is being particularly favoured. Reductionists have looked forward to bringing everything down to the level of physical explanation. Physics, they believe, as the science of fundamental particles, will show us the ultimate constituents of the universe. Physicalism, as a doctrine, insists that reality is what physics says it is. This can be softened into what physics will say one day, or would say if all possible information were in. At that point the doctrine becomes so attenuated that it loses all contact with what *we* now think. It is a narrower outlook than naturalism, which merely insists that scientific explanation is the only kind there is. This need not involve reductionism. Biological theories about organisms are not wholly reducible to biochemistry, and eventually to physics. There are truths about organisms that can be lost at a lower level. One can have a science-based thesis, which still allows for a stratification of levels of explanation.

W. V. Quine has always insisted that there is no first philosophy, no metaphysics, and makes it clear that this is a 'naturalistic claim' (1990a, p. 118). He is himself steeped in American pragmatism, as well as having been influenced by the Vienna Circle. He starts from our current theories and places great emphasis on verification. The problem, however, which he and all such philosophers have to face is how, given their acceptance of contemporary science and its methodology, they can give any justification for their commitment to physicalism and naturalism. It is one thing to hold to the latest and best supported physical theories. Given the scientific evidence available, it would be irrational in the extreme to stand outside all the controversies and make dogmatic claims based on philosophical considerations alone. Yet scientific arguments and advances must themselves always be shot through with philosophical assumptions. Because philosophy and science cannot always be separated, that does not mean that we should support science rather than philosophy. Yet that is what naturalists appear to be claiming.

Philosophy implicitly guides all scientists as they deal with reality, and sometimes, as in the dispute between Einstein and Bohr on the interpretation of quantum mechanics, differences in philosophy become of explicit importance (see Trigg, 1989, p. 165). The irony is that naturalism, the view which denies that science is indebted to philosophy, is itself a philosophical position. To hold it, we have to be able to stand back from science and talk about its scope. Naturalism seems to be a typical product of 'first philosophy'. It can hardly be deduced somehow from our practice of science, since there must always be the prior question as to whether the assumptions made by science are actually correct. Further, the belief in naturalism will on its own understanding have to be scientifically explicable. It is unclear whether we could ever be in a position to know this. Scientific explanation is of its nature predominantly causal, and does not turn to any form of rational justification. It tells us how events, including beliefs, were brought about, but not whether we are right to believe something. Yet causal explanation seems to provoke an infinite regress. Reference to each cause invites the question how *that* itself is caused. People can be caused to believe what is true. Sensory stimulation is an obvious means. Normally when I perceive the room I am sitting in I am caused to believe what is there. Indeed without such a causal foundation my beliefs would not be properly grounded. Causal chains can lead us to knowledge, but others lead us astray (Trigg, 1989,

p. 142). The trick is to know which is which and reason has a particular role in helping us to assess the character of the various causal influences to which we are subjected. Some are relevant and significant for the acquisition of knowledge. Others may not be. Naturalism is typically linked with a deterministic emphasis on causal processes, at least in connection with human knowledge of the world. Yet that means that we are always liable to assess a situation as we are caused to see it, and not necessarily in terms of its intrinsic merits. Causation is not an enemy of truth, but it is a mistake to assume that rational explanation can only be a form of causal explanation.

The Elimination of Philosophy

Unless it is accepted that human reason is capable of transcending the causal influences which undoubtedly constrain us, we can never hope to distinguish between the good and the bad. Determinism and rationality are ultimately incompatible, in that the former can never allow free reign to the latter. If everything, including our beliefs, is causally explicable, the question must always be posed whether the causes at work, described neurophysiologically or in some other scientific language, are inclining us towards the truth or away from it. The answer will itself be caused. We are therefore in principle never able to assert anything *as true* without recognizing that we are caused to assert it. We can therefore never know when we are judging correctly or whether we are only caused to believe that we are. The issue is *not* that causes always incline us to falsehood. It is that we can never be in a position to know when they do. When we are caused to have a belief, we may be lucky and be correct, and we may not be. We have no way of rationally discriminating between the two.

Does this matter? Some might doubt that it does, on the grounds that we can never question everything. We have to take some beliefs on trust. Yet it is difficult to jettison rationality consistently in favour of causal explanation. Pragmatists will say that those present beliefs which act as our starting-point must in their turn be regarded as fallible and revisable in favour of other beliefs. This might perhaps suggest a rational process of revision which may not be open to a consistent naturalist. Naturalism relies heavily on the categories of science and should perhaps regard itself as a scientific theory. Yet there still remains the question of the status of scientific theories. Some philosophers believe that all philosophical questions can be transmuted into scientific ones. Thus Patricia Churchland believes that it is wrong, according to determinism, to think that our behaviour is the outcome of compulsion and not reasoning. She asserts that what follows is that 'our reasoning and our reasoned behaviour is causally produced' (1981, p. 99). Rationality is thus explained rather than undermined. She continues: 'So far from denying that humans are purposeful and reasonable, determinism is the thesis that there is a causal network which produces such behaviour.' As we shall see, current developments in physics anyway put in question the deterministic vision of the universe, or even parts of it, as a closed and predictable system. Perhaps, it might be suggested, this is an example of science bringing about revision of its own self-understanding. It should be clear, however, that determinism is as much a metaphysical thesis as any belief in human free will. It is not a deduction from science so much as a presupposition of it, at least in some areas of science. When causal explanations are not forthcoming, it is always possible to assume that this is because of a fault on our side in that we have not been able to discover the appropriate mechanism.

Patricia Churchland, however, is not only a determinist, but a reductionist. She finds no difficulty in reformulating even basic issues concerning the status of knowledge in neurophysiological terms. It is then easy for us to suppose that science is equipped to deal with those issues. She comments that from Plato onward the fundamental epistemological question is how it is possible for us to represent reality. She says:

> Since it is the nervous system that achieves these things, the fundamental epistemological question can be reformulated thus: *How does the brain work?* Once we understand what reasoning is, we can begin to figure out what reasoning *well* is. (1987, p. 546)

There is a definite slide here from questions of justification to issues about the function of the brain. Churchland considers that she is not changing the subject, but her reformulation of the question betrays a radical switch. The original question, as she presents it, as to how we represent reality, is not as easy to deal with as she imagines. Plato would certainly not have accepted that he was in fact concerned with the working of the brain. The philosophical concern is, and always has been, not how brain processes put us in touch with reality, but how we know that it is *reality* we are representing. Once we know what reality is, no doubt we can investigate the crucial role played by the brain. Churchland, however, chooses to assume that a scientific description of the processes is also their validation. As a result, science appears to be able to deal with the prime philosophical issue of distinguishing good from bad reasoning.

Churchland is demanding a major change in our understanding of philosophy, and it is unclear whether philosophy can be left with any function at all. This is particularly illustrated in the way in which it is now being suggested that science has the capacity to solve the traditional philosophical question of the relationship between the mind and the brain. It is, however, all too apparent that those, including Churchland herself, who suggest this envisage the elimination of the mind. Neural science will replace psychology. Two neurophysiologists write: 'Our working hypothesis . . . is that levels of organization exist within the brain at which a type-to-type physical identity might be demonstrated with cognitive processes' (Changeux and Dehaene, 1989, p. 44). Such a quest for identity is not made with any thoughts that the cognitive processes will define the various levels of organization within the nervous system. Although, somewhat paradoxically, scientists may depend on a prior understanding of cognitive processes to help them in their research, once empirical correlations have been made, they seem ready to jettison their 'folk psychology' and rely on their scientific understanding of neuronal processes. They will have abandoned the very basis on which they were able to build up their theory. Thus a switch is made from the mental to the physical, the philosophical to the scientific, and the rational to the causal. Once correlations in the brain are established, talk of the contents of mind is changed to reference to 'transient physical states of neuronal networks' (Changeux and Dehaene, 1989, p. 93).

The relationship of the mind to the brain is a philosophical problem of venerable age. The modern breed of neurophysiologist, cheered on by some philosophers, goes further than simply reproducing a particular philosophical answer to the problem by advocating a materialist view of the mind. It is claimed that the problem itself has been eliminated. Science can show us, we are told, that no philosophical issue remains. We have all the information we need once we see how the brain works in building up our view of the world. There is, it seems, a 'neural architecture underlying the faculty of reason' (Changeux and Dehaene, 1989, p. 70). The science of the brain can show us the working of our reason. Rationality is a product of the brain, and hence itself susceptible to scientific examination and explanation. Yet if human science is itself the outworking of human reason, then it must be a question as to how far it can itself explain, let alone explain away, the fact of human rationality. If science is asserting certain things as true, and demands attention because it provides us with knowledge, these are intrinsically rational claims. When science attempts to describe our reasoning as a certain form of neural function it is making claims on our reason. This point might appear to be accommodated by reference to different levels of organization within the brain. The point is though that however our brain works, appeals to reasons for belief and attempts at justification can never themselves be part of a scientific theory. Science depends on reason. It presupposes it and cannot itself somehow dispose of it with a reductionist programme. What makes us believe and whether we are right to believe are questions that can never be assimilated. Even if it is shown that certain causes of belief are relevant pathways to truth, this presupposes a prior, and rationally based, understanding of where the truth lies.

This form of argument is the antithesis of the naturalist position. According to one form of naturalism, if something is accepted in science, that is a reason for accepting it is philosophy. That is a very weak type of naturalism, which retains the language of reason and the distinction between science and philosophy. Much more threatening is the outlook which refuses to allow philosophy a status distinct from that of science. Even Richard Rorty signs up for the naturalist cause. This is curious, as a relativist outlook sits uneasily with the view that science shows us the actual causal processes inherent in the world. If science is only one conceptual scheme amongst others, it has no apparent right to inspire a naturalist commitment which extols the superiority of scientific over other forms of explanation. Rorty defines naturalism as 'the view that *anything* might have been otherwise, that there can be no conditionless conditions' (1991b, p. 55). The idea of contingency certainly underlies his post-modernist approach, and espousal of science can certainly produce an awareness of the utter contingency of things. Yet why should this challenge an idea of human reason? Only if reason is conceived of as simply the intuition of necessary, timeless truth, as in mathematics, does this make sense. Yet there may well be other kinds of truth. Reason may still have a place even if the whole distinction between necessary and contingent, or analytic and synthetic, is rejected. There appears, too, to be a slide in Rorty from the affirmation of contingency to the denial of objectivity. Yet it is surely an objective truth about the real world that it is contingent. Rorty asserts: 'Naturalists believe that all explanation is causal explanation of the actual, and that there is no such thing as a noncausal condition of possibility.' It follows that philosophy seen as the quest, as Rorty puts it, 'for truths whose truth requires no explanation' is opposed to naturalism and should be discarded. Needless to say, it seems odd that philosophy should be jettisoned in favour of a view, such as naturalism, which itself appears highly philosophical and even to be supported by philosophical argument.

One of the key elements in a naturalist outlook is that there is no sharp distinction between different kinds of organism reacting to their environment. Humans cannot set themselves fundamentally apart from the rest of the biological world in virtue of their much vaunted rationality. The amoeba adjusting itself to changed water temperature is no different in principle from the apparently rational activities of humans. Rorty himself sees no break in a hierarchy stretching from the amoeba at the bottom to 'bees dancing and chess players checkmating in the middle, and people fomenting scientific, artistic and political revolutions at the top' (1991a, p. 109). This clearly fits in with Darwinian theory. It is apparent, however, that in using naturalism as a means of defeating metaphysics, Rorty is making claims as to how things are. Naturalism holds that it is false that minds can have any form of existence apart from the brain. The emphasis, too, on organisms adjusting to their surroundings only makes sense within a philosophical outlook that recognizes the reality of the organism, the environment, and indeed of the causal connections allegedly holding between the two. Naturalism tells us about the nature of the world and our place in it. It cannot be combined with any view, such as Rorty's, which tells us that there is no such thing as 'the nature of the world'.

Naturalized Epistemology

Naturalism is not just a theory about reality, but also makes claims about what we can know. The whole idea of a 'naturalized epistemology' has become very influential, particularly because of the work of Quine. In its strongest form, as one might suspect, it makes epistemology a branch of science. It is left to the latter to discover how we are put in touch with the world. The norms of epistemology are replaced with empirical facts about the working of the brain. There is no question of epistemology showing us the norms for the selection of beliefs. We start with science as the authority, on the grounds, presumably, that contemporary science is our best bet for obtaining knowledge. There must therefore be a concealed epistemology, which is used to justify contemporary science, underlying even the boldest naturalism. Perhaps because of that, some who wish to embrace a naturalized epistemology are more cautious, and either use the discoveries of science to give flesh to their epistemology, or simply to make the judgement that scientific method is the only reliable way of obtaining knowledge. That may be controversial but it is certainly a proper judgement in the context of epistemology.

The proven methods of science must be relevant to us in our

quest for knowledge. At its weakest, naturalized epistemology warns philosophers not to ignore the findings of science, for example in considering facts about perception and its role in our acquiring an understanding of our surroundings. At its strongest however, naturalism in general and the idea of a naturalized epistemology in particular undermine epistemology, because they undermine reason. We cannot assess arguments or judge the validity of theories since this suggests the ability to distinguish the good from the bad, the true from the false. Causal explanation replaces justification. Naturalism's attack on the possibility of any metaphysics has the effect of making epistemology itself impossible. All we can do is see why people adopt certain beliefs and not whether they have good reasons.

Quine claims that 'naturalism looks only to natural science, however fallible, for an account of what there is' (1992, p. 9). This appears to imply that we are in a position to judge the efficacy of science from outside, but Quine's denial of metaphysics means that he is actually reaffirming a commitment to science. Only questions already couched in the language of science are of interest to him. It is a matter of seeing how scientific theory is related 'to the triggering of our neuroreceptors by the external world' (p. 8). We are dependent, it seems, on what natural science tells us about our cognitive access to the world. Objects cannot be seen as antecedently existing, but have to be understood from our point of view. It becomes a question as to how we can build up the concept of an object from 'an amorphous welter of neural input'. Quine says: 'Even the notion of a cat, let alone a class or number, is a human artefact, rooted in human predisposition and cultural tradition' (p. 6). We do not see 'what is there' but are dependent on various neural mechanisms in building up a picture of the world. We gradually get further from the sensory periphery until eventually, as part of the same continuous process, we can even refer to the unobservable entities of theoretical physics.

Quine is quite prepared to make categorical claims but they are always made from the standpoint of a particular conceptual scheme. Science is about what it claims to be about, but this not because it is in touch with an independent reality. Such a claim makes no sense for Quine. The reality is not of some metaphysical kind existing whatever scientists may happen to think. It is posited by science, and if science is revised our ideas of reality will have to change. Science is our only means of access to reality. Reality is what science says it is, and even our most basic categories such as that of an object are human ones. We cannot check our categories against reality, since we only view 'the real' by means of our categories. Science is a human construction itself, and yet also shows us how our concepts can be built up. The quest for reality apart from science is the quest, Quine believes, for knowledge after we have discarded our only means for obtaining it: 'It is like asking how long the Nile really is, apart from parochial matters of miles or meters. Positivists were right in branding such metaphysics as meaningless' (1992, p. 9). Quine parts company with positivists who wish to tie meaning too closely to sensory stimulation. He wishes to follow wherever science leads and not impose a philosophical strait-jacket. His naturalist commitment to science paradoxically means that no philosophical doctrine can be imposed from outside, even one intended to justify science as our only means of knowledge. If physics wishes to refer to quarks, then we are right to refer to them, even if they are unobservable: 'The world is as natural science says it is, insofar as natural science is right and our judgement as to whether it is right, tentative always, is answerable to the experimental testing of predictions' (p. 9).

We devise and then revise theories 'in the light of physical impacts on our physical surfaces'. In true pragmatist fashion, Quine holds that we cannot stand outside our current beliefs. We can change them, but cannot appeal to metaphysics as a source of justification. We start with sensation, or at least the neurophysiological underpinning of sensation, and must pull ourselves up by our bootstraps from there on. He is not basing any epistemology on sense-experience, still less looking for a foundation for certainty. He starts from within science and his references to neural receptors underline his naturalist position. He claims: 'It is a finding of natural science itself, however fallible, that our information about the world comes only through impacts on our sensory receptors' (1990b, p. 19). He adds that the point is normative, 'warning us against telepaths and sooth-sayers'.

The contradictions inherent in the idea of a naturalized epis-temology come to the fore again at this point. Any epistemology ideally recommends procedures which can be reliable guides to truth. It shows us which tactics are good and which are useless in the pursuit of truth. It must, above all, be normative, providing us with justification for certain types of belief, and warning us against others. For example, believing something because we want it to be true may be a common human failing, but it is not recommended from an epistemological point of view. Believing something because the best scientific evidence suggests it is right is much more sensible. This is not simply because 'we' think like that, but because the methods of science have been tried and tested according to criteria that transcend scientific method. The practice of science can be justified on rational grounds and not merely in its own terms. Indeed science on its own cannot be in the business of justification at all, since it is intended to be descriptive. This is the tension at the heart of naturalized epis-temology. It may show how we do build up our picture of the world, but the shadow of traditional concerns with justification still hangs over it. The normative and the rational sit uneasily with an austere science which is apparently purged of all meta-physical assumptions. Neural stimulations, by themselves, seem to provide a poor basis for our building up knowledge about the world. There is an obvious jump from descriptions of them to reference to justified belief. They may be causally involved in our interaction with the external world, but there is still the problem as to how they can underwrite our claims to knowledge.

Quine agrees with empiricists concerning the importance of the problem of the relation of science to sensory data (1990b, p. 19). He has reinterpreted such data in terms wholly acceptable within a scientific theory, indeed imbued with theory. In so doing, he has removed the traditional foundations of science, the sense-data which appeared to positivists to provide an incorrigible basis for knowledge. The neural data Quine appeals to have no such epistemological status. They are the product of theory and not its justification. Neural stimulation can hardly act as a constraint on our knowledge and yet, intriguingly, Quine is still enough of an empiricist to believe it should, since he wants to rule out sooth-sayers and the like. Yet this serves to reinforce the insistent question why we should abide by the norms of science. If it has certain procedures and rules out others, if it places a premium on confirmation obtained through the sensory organs, and dis-counts purported knowledge from other sources, we can no longer simply say that science is where we start. There seem to be al-ternative starting-places. What for instance, is wrong with astro-logy? Many in our society would see nothing wrong at all with it. Why should we not start from where *they* are? We must have reasons for opting for a scientific outlook, when so many alter-native procedures are all too obviously on offer. Norms exclude rivals, and unless we consciously embrace some, we will face paralysis when confronted with a major challenge to our existing practices. It is useful to have an answer to the question why one should go on doing what one is doing. It is a question we may on occasion even ask ourselves.

'The Science Game'

In talking of predictions as the checkpoints of science, Quine appears willing to talk of science as a language-game, in Wittgenstein's sense, contrasting it with other language-games such as fiction and poetry (1990b, p. 20). He claims that predic-tion is not its main purpose. Instead, he says 'nowadays the over-whelming purposes of the science game are technology and understanding'. This reference to games is ominous, though not surprising, since it encourages the view that description of the rules of the game and not its justification is all we can hope for. One need not like poetry, or even read fiction, just as one need not play golf. Yet for a naturalist, science does not seem to be in this category of a hobby or optional interest. Its practice constitutes the naturalist's approach to the world, and it purports to rule out irrational practices and such mistaken theories as 'folk psy-chology'. Quine's reference to games suggests a certain lack of confidence in science as our conceptual scheme. The very conception of something as a language-game involves a distancing of ourselves from it, and a recognition that it is merely one form of human practice. This sits uneasily with the naturalist emphasis on the physical. Indeed naturalism can easily also involve physicalism, claiming that reality is as the fundamental science of physics sees it. Not all naturalists would accept the reduction endemic in such a position. In the case of Quine, his desire to rule out telepathy and soothsaying stems from an assuredly physicalist outlook. He has always espoused physicalism. Yet he now claims that 'the science game is not committed to the physical, whatever that means' (1990b, p. 20). It turns out that because he views empiricism as itself empirically demonstrated by science, he is willing to envisage the possibility that telepathy and clairvoyance are, as he terms it, 'scientific options'. The empiricist belief that knowledge comes only through the senses comes from science and as Quine says 'science is fallible and corrigible'. He regards sensory prediction as the checkpoint of science, and if 'telepathy or revelation' provided extra knowledge, our present scientific beliefs would have to change. He continues: 'In that extremity, it might indeed be well to modify the game itself and take on as further checkpoints the predicting of telepathic and divine input as well as of sensory input' (1990b, p. 21).

This is an example of the baleful effects of the meeting between pragmatist, fallibilist ideas about science and Wittgensteinian views about language-games. Pragmatism must always be vulnerable because of its repudiation of any idea of a grounding for science. The very idea of warranted assertibility raises the question of the source of the warrant. A strong empiricist position at least gives guidance as to when theories might be given up or modified. Once it is asserted that it is a scientific rather than a philosophical position, it can apparently be overthrown in the light of countervailing evidence. This is an inconvenient fact in that this is as much a corroboration of empiricism as its defeat. The reliance on empirical evidence would seem to lie at the heart of empiricism. Yet the catch is deciding what counts as proper empirical evidence. The temptation for a strict empiricist will always be to say that revelation, telepathy and the like can never be proper sources. Empiricism, as Quine stresses, is precisely the view that all knowledge comes from the senses, and such a position need not be vulnerable when alleged knowledge is obtained elsewhere. It is too easy to rule the new source out as beyond the scope of science, and therefore not to be taken seriously.

The very fact that Quine admits that predicted sensation may not in the end be enough as a checkpoint, and that the science game may have to be modified, shows an incoherence lurking behind his willingness to give up empiricism. He is presumably ready to jettison naturalism too, since empiricism is regarded by him as the 'crowning norm of naturalized epistemology'. We begin with a position that upholds empiricism and provides warnings against telepaths and soothsayers, and end with one that has given up empiricism because of what the telepaths have told us. We begin with an absolute commitment to the presuppositions of present-day science, including a strong naturalism, and end with a weak idea of science as one language-game amongst others. Moreover it is one that can change through time in a fundamental way. Perhaps this happens because we discover that the world is not as we thought it was. The question is *how* we can discover this. The clear, if constricting, guidelines of empiricism are discarded in the light of evidence that would not be accepted as such by a genuine empiricist.

This curious position has been reached because of the naturalist's contention that metaphysics, and even a genuinely normative epistemology, have no place in our reasoning. Philosophy cannot then lay down ground-rules about how the world is to be conceived, or about what counts as good evidence for belief. If our starting-point has to be science and if scientific discoveries were to challenge the materialistic assumptions on which science is allegedly based, it is not surprising that we get caught in a certain incoherence. Science seems to subvert itself. What is necessary is a recognition that even materialists uphold a definite metaphysical position, which can only be supported or condemned on rational grounds that may not themselves be narrowly scientific. Whether the admission of such metaphysics supports or destroys naturalism is another question. If naturalism is intrinsically anti-metaphysical, the admission that it cannot itself do without a metaphysical grounding, even of an austere kind, may be fatal. What may partly be at issue is the precise relevance of empirical 'evidence'. Empiricism certainly has had an extremely narrow view of what can constitute such evidence. Its emphasis on sensory experience peremptorily rules out other forms of alleged experience – for instance, that claimed by mystics. Whatever its possible defects, however, it is clearly a philosophical theory about the epistemological status of sense experience. The relationship of empirical evidence to a particular scientific theory is a secondary issue, which is the proper concern of our epistemology. Yet naturalism deliberately blurs the empirical and the philosophical. It is fallacious to suppose that just because a theory is about the empirical world, and our understanding of it, that it must itself be an empirical theory. Philosophy can and must deal with the nature and status of the world, with how we can obtain knowledge of it. This ability provides the framework in which science can operate. It is not itself a product of science.

The project of a naturalized epistemology will either be a philosophical theory smuggling epistemological norms into science, or a scientific one which can then make no rational appeal to justification. Quine's tendency to treat 'sensory stimulation' as the basic datum which then provides evidence for belief, shows that he has not really produced a scientific account at all. He is still bound by epistemological constraints. Quine in fact explicitly refers to 'the relation of evidential support', which he locates within what he terms 'the baffling tangle of relations between our sensory stimulation and our scientific theory of the world' (1990b, p. 1). He thinks that this can be separated from neurology, psychology and other disciplines, which means that it must be within the province of an epistemology which is irreducible to science. 'Evidence' is not a scientific notion, but is firmly placed in the context of assessment and judgement, and not of mere description. In fact science has to be seen as itself an expression of human rationality, rather than rationality being redescribed in exclusively scientific terms.

What are the respective roles of philosophy and science? As we have already noted, Quine's basic position, like that of so many philosophers, has been that 'there is no external vantage point, no first philosophy' (1969, p. 127). Philosophy is continuous with science. Yet he still seems to allow a residual place for an epistemological consideration of evidence, rather than a mere recital of causes. A consistent naturalist, however, should not even conceive of referring to evidence and justification, but should be willing merely to follow the apparent discoveries of science. Anything else would suggest the ability to stand apart from science and appeal to standards of rationality. Modern science may well throw light even on human psychology and the working of human rational processes. Naturalism, however, is more than an exhortation to take such findings seriously. To say that there is no external vantage point, means for the naturalist that we must rely absolutely on science. Philosophy, viewed as a discipline in any way distinct from science and concerned with the nature of reality and an ability to reason about it, is an illusion. Naturalism may, of course, accept that it is itself a metaphysical thesis saying that reality is such that only science is capable of properly discovering its characteristics. Since, however, naturalism is intended as a denial of metaphysics, it is ill-advised to retreat to any metaphysical high ground.

Scientific Realism

One writer on realism in science suggests that its point is to support a 'naturalistic world view' (Hooker, 1987, p. 25). He defines the latter as the view according to which 'objects, geometry, man, the universe at large are taken seriously as science knows them.' At first sight, the view that reality is independent of our conceptions of it does not sit well with the view that reality is defined by science. Science is the product of human minds, so it could be argued that naturalism actually undermines realism by making tacit reference to human capabilities. The thesis of scientific realism makes the connection. It is however not so much a theory about the status of science as a thesis about the meaning of scientific claims, suggesting that they are intended as a literal description of the physical world. Scientific realists would add that such descriptions are very often successful in referring to the real world. Indeed they tend to pin their faith on the correctness of much contemporary science, holding that much of current theory is at least approximately true. They would further suggest that this fact provides the best explanation for the success of science. Otherwise, it is alleged, it would be a miracle that we can make successful predictions. A scientific realist would argue that mere pragmatism is not enough, because we need some rational underpinning for the fact that we have confidence in our predictions. That can only be given, it is claimed, by an appeal to the nature of reality as demonstrated to us in science. The argument is that there is good reason to believe in the existence of entities which have been posited by theories which have continued to generate reliable predictions and to give us increased control over our surroundings. The best explanation for that kind of success is that we have gained an insight into the structures of the real world.

Scientific realism starts with science and then aims to explain its apparent success. By definition, therefore, it has to limit reality to what is described by science, and to what is within the scope of scientific method. Indeed for many, the whole point of scientific realism is to rule out the possibility of ghosts and similar entities. A more metaphysical form of realism would start with reality, and not with science. Because of this it could be open to other forms of knowledge, and would also be reluctant to take scientific success at face value. Instead of inferring reality from our apparent success, it would consider our success to be such as would have to be properly grounded in reality. Just because there is no simple way of proving this, a logical gap is opened up between reality and our theories about it. As a result, a metaphysical realism can always be criticized for allowing the spectre of scepticism to haunt us. Scientific realism, on the other hand is not in the least sceptical and can be criticized for being too ready to accept current science as true. The history of science and its continued change can hardly give anyone confidence that present views are going to be sacrosanct. Indeed scientific realism is open to disproof if present-day science is one day demonstrated to be seriously mistaken. It is in fact more of a thesis about science than about reality. It is hardly surprising that this kind of realism can be the close ally of naturalism.

A realism that is allied with science typically explains the relationship between the subject and object of knowledge as a causal one. It eschews reference to conceptual or logical necessity. C. A. Hooker makes this point when he links realism and naturalism:

> The realist can allow very strong causal interactions between mind and reality, for that precisely still involves a reality whose basic character is mind-independent. This picture of humans is supported by the naturalistic, evolutionary conception of ourselves from science. (1987, p. 256)

Any argument about the relationship of philosophy and science, and more particularly metaphysics and science, often poses a stark choice between the causal connections discovered by science, and the logical ones revealed by philosophy. There is a clear split between the contingent and the necessary with science on the one side and philosophy on the other. Yet once it is emphasized that the rational basis of science lies in the world, this distinction can become blurred. Assertions can be made about the nature of contingent reality, and, for instance, its orderliness, which precede science rather than being dictated by it. Naturalism, however, tries to avoid all this even when it embraces realism. It makes causal relationships and not rational ones the basis of knowledge. Indeed it would deny the distinction, by stressing not just that there are objects of knowledge, but that science can explain our connection with them. Since the theory of evolution purports to show on the grand scale how our biological species has developed in the world, a naturalistic epistemology may turn to this.

Indeed John Searle refers to the atomic theory of matter and the evolutionary theory of biology as in large part 'constitutive of the modern world view' (1992, p. 86). Although he wishes to demonstrate the importance of consciousness, he firmly believes that for it to be placed within our understanding of the world, it has to be situated with respect to these two theories. This itself is to concede everything to science. Rightly or wrongly, it is to say that our present most firmly held scientific theories provide the context in which we work out our philosophy. Searle comments about the status of these two theories: 'Like any other theory, they might be refuted by further investigation, but at present the evidence is so overwhelming that they are not simply up for grabs'. Nevertheless the result of this approach is to accept without question the modern world view without attempting to give it any rational basis. Our philosophy seems to be built on a sociological fact about which views seem to have gained acceptance. If it is maintained that the views in question are the most rationally established that we have, this might be true, but it begs the question as to how we know what is rationally established. 'Because science tells us so' is not an adequate answer for anyone asking about the basis of naturalism.

Naturalism holds that the world of which we are a part is a unity from which we cannot detach ourselves. The resources of Darwinian theory can be utilized to explain our knowledge in evolutionary terms, so that natural selection appears to be the driving force behind the development of the human mind and even of particular beliefs. The argument is that it is to our evolutionary advantage to be able to recognize features in our environment. Unless our ancestors had become fitted to the world they would never have survived or produced offspring. Knowledge can promote survival. This can certainly explain ordinary empirical knowledge, but it is doubtful if it can directly explain the growth of scientific knowledge. Apart from the fact that science may not actually promote our survival, it is the product of a brain that evolved long before the advent of scientific reasoning.

Human rationality may have been produced by the process of natural selection, but it is unlikely that many of our beliefs, particularly of a theoretical kind, are dictated by our genes, or even by our genes in interaction with our environment. Genes must, in order to spread, have encouraged behaviour conducive to reproductive success. Yet the pursuit of reproductive advantage and the desire for truth are not necessarily the same. Acquiring scientific knowledge could easily prove deleterious. When science gives us the means of destroying ourselves and our planet, the pursuit of truth might seem highly risky from an evolutionary point of view. Similarly, false beliefs may actually prove to be an advantage. Sociobiologists who are atheist and sceptical about the possibility of moral knowledge can still see great reproductive advantage in adherence to religious belief and the consequent felt obligation to further the interests of others. The argument is that a certain level of altruism helps a community to function and its members to flourish, even if the religious foundations of altruism are illusory. It may even seem that it is important from an evolutionary point of view that we are deceived as to the truth of the matter. Some argue that natural selection has, through the development of certain physiological processes, accomplished precisely this. We are told that 'human beings function better if they are deceived by their genes into thinking that there is a disinterested objective morality binding upon them, which all should obey' (Ruse and Wilson, 1986, p. 179). Such a naturalistic ethics graphically illustrates how the interests of truth and of natural selection need not coincide and may actually conflict.

One of the undoubted advantages possessed by the human race is its remarkable flexibility and its ability to adapt to many different, and even rapidly changing, environments. The idea that we are born with a tendency to acquire many rigid beliefs would not seem to fit with this picture. It is not surprising that many epistemologists have preferred to take evolution as an analogy, rather than as an actual mechanism in the development of knowledge. Karl Popper, indeed, appears to drift rapidly from one to the other. While insisting that 'from the amoeba to Einstein the growth of knowledge is always the same', he writes: 'While animal knowledge and pre-scientific knowledge grow mainly through the elimination of those holding the unfit hypotheses, scientific criticism often makes our theories perish in our stead, eliminating our mistaken beliefs before such beliefs lead to our own elimination' (1972, p. 261).

The process by which the fittest organism survives is very different from the way in which scientific theories are rationally assessed. The one appears to be a blind process when chance rules, whereas the testing of new theories is very far from being the accidental product of the physical world. The search for new theories is itself not random. For one thing, the number of possible ones would always be such that we have to select what would appear to be the most promising in order to test them. The search for truth and the elimination of falsehood is a highly purposive exercise which does not reflect the process of natural selection as viewed by Darwinism. Reason and truth provide a different framework for science. Evolutionary epistemology, as a distinctive programme, only makes sense as a strong form of naturalism, which equates rational with causal explanation, and truth with questions of reproductive advantage. In the words of Hooker, we need a naturalistic, foundationless epistemology such that 'epistemology theory is strongly determined by our scientific views of the world and our place in it' (1987, p. 22). Once again we find that epistemology is made to follow science and not precede it. This is only reasonable if we are already assured that science is a source of knowledge, and further that it is our only one. Yet evolution cannot give us such an assurance, and even if it did, we could not be sure that it was reliable. It could be the kind of illusion that we are told moral knowledge is.

The theory of evolution itself claims truth and is put forward for reasons. It purports to be about the real world and to deal with actual mechanisms at work there. Evolutionary epistemology, therefore, depends on the prior adoption of the theory. There is the constant danger that our epistemology is governed totally by whatever happens to be the currently accepted theories in science. Yet we have already adopted those theories as the result of some previous epistemology, held implicitly or explicitly. Too many philosophers are willing to be governed by our best contemporary scientific theories, even if they recognize their fallibility. Yet how do we know which are our 'best' theories, other than because people happen to believe them? It seems as if all the most crucial decisions are taken *before* we produce an epistemology. It is not very comforting if they are being taken arbitrarily or because of social pressures. If, on the other hand, they are being taken for reasons, this suggests that rationality, and not just the scientific discovery of evolutionary mechanisms, lies at the root of our understanding of our place in the world. Any naturalistic theory has to justify itself. Not least among its tasks is to explain how, if it is true, we are in a position to see its truth.

5
A God's-Eye View

An Absolute Conception of Reality

The possibility of giving an external justification of a body of belief and practice, such as science, is highly controversial. We have already seen how pragmatists, Wittgensteinians and others all agree in challenging the idea of a rationality that is detachable in any way from its context. They would claim instead that the reasons we give can only be 'internal', i.e. that they are generated from what we already do. It would follow that a rational justification of a whole practice would be impossible. We could never, it is alleged, achieve the neutrality between views that might appear to be necessary. Doing so would imply that we could put ourselves in the position of God, seeing and knowing everything. What we must do, it is agreed, is offer arguments which already take for granted where we are. We cannot detach ourselves from everything so as to stand nowhere. We cannot stand apart from science and reason about it. That was the pragmatist position, and it is also in effect the position of the naturalists who are convinced that scientific method can be the only path to truth. It is important, therefore, to question how relevant the notion of a God's-eye view might be. The very impossibility of reaching such a position may suggest the impossibility of a rationality that seeks to justify our most basic practices. Yet rationality should be concerned with the character of reality, and not necessarily with arriving at a particular viewpoint. We need now to question the alleged link between reality and the possibility of having an absolute conception of the world.

Scientific beliefs develop and change with alarming rapidity. We remarked in our discussion of scientific realism how the history of science hardly suggests that present scientific beliefs are sacrosanct. The contrast between what we think we know at any given time and what is actually the case must be very marked. We are, however, never in a position to notice it, since all we can do is to form conclusions on the best evidence available at any given time. There will always be room for controversy as to what exactly is the 'best' evidence, but no one can be in any doubt that later generations will look back on us and see the partial nature of our knowledge. However used we are to the notion of scientific progress, it implies that even we may be seriously mistaken in some of our most cherished views. What there is should not be too closely identified with any one conception of it. Peirce's idea of an ultimate agreement between scientists is an effort to tie down conceptions of reality with what an identifiable group of people might believe. Yet the group is too hypothetical. The danger is always that the concept of those with complete knowledge is as empty as that of a reality waiting to be known.

The concept of reality seems to demand as a corollary the idea of an absolute conception of the world. All available conceptions seem too limited and partial. Even if what we say is true, we are still asserting it from a particular standpoint. Bernard Williams, writing about Descartes' attempts to indulge in 'pure enquiry', refers to the possibility of a determinate picture of the world, 'independent of any knowledge or representation in thought' (1978, p. 65). He makes the obvious point that this picture must still be ours, presented from our point of view. He says of it that it can itself be seen as 'only one particular representation of it, our own.' It will appear that 'we have no independent point of leverage for raising this into the absolute representation of reality'. We thus appear to face a choice between a conception which may be regulative but which we certainly cannot lay hold of, or one that is all too clearly ours. Indeed the idea of an absolute conception is itself an unstable one. If a conception is absolute, there is presumably no room for doubt, partial knowledge or error. It carries with it the idea of a full and complete vision of reality. Yet a conception must be had by someone. The very demand for absoluteness seems to deny the possibility of ordinary humans situated in particular cultures ever attaining it. This is underlined when the historical nature of our understanding is stressed by Rorty and others.

There is, though, a pronounced strain in Western philosophy which encourages us to believe that we are capable of transcending our particular limitations. Plato rooted knowledge in the relation of the soul to a reality which lay beyond the confines of this life. This view passed into Christianity, and St Paul was perhaps echoing Plato's contrast between image and reality when he said: 'At present we see only puzzling reflections in a mirror, but one day we shall see face to face. My knowledge now is partial; then it will be whole, like God's knowledge of me' (1 Corinthians 13). He was contrasting the limited knowledge we have in this life with the complete knowledge to which we could attain in the presence of God. Religious and mystical ideas become linked with the goal of absolute knowledge. Certainly the idea that God must act as the fountain of knowledge and the guarantor of truth runs deep in Western philosophy. Such a God may sometimes seem nothing more than a rational necessity, there to provide the ultimate grounding or foundation for belief.

Descartes believed that he had proved the existence of God and hence could rely on Him as the guarantor of the existence of an objective reality. He thought that our judgements would themselves be generally reliable, since God is not a deceiver. We could thus place some confidence in 'the light of nature', or the faculty of knowledge which God has given us (Descartes, 1911, pp. 231 and 211). As Williams comments: 'One might say that what God has given us, according to Descartes, is an insight into the nature of the world as it seems to God, and the world as it seems to God must be the world as it really is' (1985, p. 139).

This type of philosophy grounds our understanding of reality by reference to God. We can trust our power of reasoning since God has given it to us. Any adequate account of reason and reality must show how we can have the conception of the world we do. It is not enough for us to be able to provide an understanding of the nature of the world, unless we can at the same time demonstrate why we can attain such an understanding. Descartes' reliance on God as the guarantor of reason circumvents this objection. He holds that the very fact that we can have a conception of God as a Perfect Being means that He must exist. That is the consequence of the insistence through the ontological argument that God's existence is actually implied in the idea we have of Him. Given that we can conceive of God, it follows that our belief in a system of thought founded on Him is going itself to be reliable and not merely an arbitrary starting point. Such a conjuring of existence out of a human concept may seem highly suspicious. Nevertheless the notion of God as instantiating an absolute conception of reality provides us with a fixed standard against which to measure our partial knowledge. The omniscience of God seems to define what knowledge is.

This emphasis on the fact of God's knowledge as the grounding of knowledge, instead of God's Being as the grounding of existence, perhaps suggests an excessive preoccupation with the nature of knowledge rather than the nature of reality. In fact the question must arise if this course is pursued whether reality is as it is because of God's absolute knowledge of it. Indeed does God's understanding of Himself constitute His reality? This is the path to idealism and one which refuses to accept the brute fact of existence. The insistence on relating reality to a particular perspective, or, in the case of an absolute conception, to the transcending of perspectives, does not do justice to reality. It has changed the subject to knowledge. The very idea of an absolute conception depends on the notion of a conception that is somehow not a conception, of a view that is not exactly a view. It is an attempt to refer to reality whilst still mysteriously appealing to some ideal observer, who manages to observe whilst avoiding all the disadvantages of observation. What is real is then seen but not from a particular perspective. It is known, but in a way that transcends all the normal possibilities of knowledge.

The whole point of talking about reality is to refer to something as it is independent of all conceptions. Admittedly, there is a sense in which we have to have a conception of what is independent of all conceptions. This, though, is part of the general problem of whether human reason can transcend its own inherent limitations. Can we rationally grasp what is real? This is not the same question as whether the notion of an absolute conception is meaningful. The latter refers to the possibility of an all-seeing eye, a viewpoint that is not limited. The former deals with the issue of whether we can have dealings with whatever exists, as it is in itself. The notion of an absolute conception is connected with that of reality since by definition it mirrors it. The one, though, is an epistemological question, the other ontological. The argument that one cannot refer to reality without involving some conception of it, can easily divert attention from the nature of existence to the difficulty of attaining more then a partial conception. The emphasis has moved to the nature of our understanding, and away from what we were trying to understand. Yet if anything exists, it is as at is. Its existence may depend on God, if He exists, since He has created and continues to sustain everything. It is in no sense logically dependent on His knowledge of it. A firm distinction must be drawn between the causal dependence of all creation on the Creator, and the logical independence of what is real from the knower.

The blurring of the boundaries between the epistemological and the ontological is inevitable once the idea of an absolute conception of reality is given prominence. This becomes all the clearer in the work of Bernard Williams, who links the notion with the possibility of an ideal convergence of belief by scientific investigators. He uses it to emphasize a distinction between the world independent of our experience and the world as it seems to us. The former is the world open to science, even the science of aliens. An absolute conception thus turns out to be something not so very different from Peirce's view of a final agreement between scientists. The omniscience of God has been replaced by monopoly claims to truth for science. Indeed, Williams draws an explicit contrast between the possibilities of convergence of belief inherent in science, with the impossibility of anything similar happening in ethics (1985, p. 136). Reality is, in this way, indissolubly linked with the possibility of an absolute conception, and the latter in turn seems to become nothing more than the possibility of convergent belief. One does not have to stretch this argument very far to conclude that reality is what scientists can agree about, even if only in principle. Yet, giving science this kind of primacy is to give a privileged status to what may be only one method of gaining knowledge. It not only links ontology to epistemology, but makes the criteria of contemporary science the only acceptable ones for epistemology.

It is important to distinguish questions of reality from those of knowledge (even perfect knowledge), since many metaphysicians have wanted to ground reason in reality. This is not the same as basing it on knowledge. For one thing, questions about knowledge can easily be transmuted into sociological questions about the standards of justification operative in a society. Reality cannot be relativized to a society without being denied. The idea of objective truth, detached from issues about the character of possible knowers, has played a dominant role since Plato. He tried to ground our intuitions about mathematical truth and about what is just and good in the very structure of reality. Yet attempts like his often founder partly in the stubborn fact of the varying nature of human beliefs. The latter often conflict, and it is no wonder that people often hanker after some arbiter who can judge between them. The disagreements between scientists are not so dissimilar from wider controversies in society. The existence of varying conceptions of what is true runs parallel with that of varying conceptions about what is good. Scientists may agree about methodology and how, in principle, such disputes could be settled. Yet this does not in fact seem to help very much. Of course those who disagree about what is good must face a more immediate problem, in that in a pluralist society they have to live with each other. Beyond a certain point disagreement may destroy the very basis on which any community can exist. A liberal society must have some foundation, even if it tolerates considerable variation of belief.

The Pursuit of Neutrality

Liberals often argue that there is no neutral ground on which anyone can stand to adjudicate between rival understandings.

There is, it seems, no prospect of attaining a God's-eye view. The idea is that we are trapped within a particular perspective, and no one can step outside all possible opinion to get on to neutral ground. Yet this is to suggest that adjudication between conflicting positions somehow involves taking up the stance of the neutral umpire, rather than of a player in the game. It is to suggest that neutrality between positions might somehow be the same as arriving at truth. The feeling that our positions at best give us a partial view of things might encourage the notion that we need someone who is impartial. Yet the opposite of partial in this context is 'complete'. Detachment from all possible views may not be the same thing as the possession of complete truth. God's position is to be envied because of the omniscience it implies, not because of its lack of engagement with any given view. Neutrality is not necessarily a guide to truth. An ability to recognize truth may involve not taking our present conceptual scheme for granted. We must be able to stand back from it, compare it with others and question how far it is true. That is very different from opting out of all possible schemes. A God who is concerned with truth is distinct from an ideal umpire trying to keep the peace between competing interests without becoming too identified with any of them.

When faced with the needs of a society where there is very little agreement on important matters, particularly in the sphere of morality and religion, some liberal thinkers about the nature of justice, such as John Rawls, have put forward theories as to how much disagreement can be contained within a state. They deliberately avoid trying to settle substantive issues in morality, or to opt for one religious view rather than another. The state, they believe should not be involved in such matters, however important they might be in determining the life of a community. Instead they propose what is essentially a political solution whereby differences are tolerated, and individual freedom maximized. Liberalism is in effect constituted by the pursuit of such freedom. The apparatus of the state, the liberal believes, should be used to act as a neutral method of umpiring between competing opinions. The point is not to adjudicate which is right, but to ensure that each can be held without being hindered by another.

The idea of some celestial umpire upholding liberal values on analogy with the laws of a liberal state is not a good example of what a God's-eye view might consist in. Complete knowledge as compared with our partial knowledge, or total ignorance, may be very desirable, but the ability it assumes to see things as they are is very far from a disinterested detachment from all possible perspectives. Reality itself is not neutral between all possible views. Some may see more of the truth then others.

The urge to find neutral ground is like Archimedes' ancient boast that if he could find a place to stand, he could move the earth. We want to abstract ourselves from our context and get outside all competing schemes. We want to transcend all possible vantage points, unaware that we are going to end up at just another vantage point. Indeed it could be argued that the liberal vision of a just society, where individual freedom and equality are rated more highly than particular conceptions of what makes a good society, is in fact itself a specific view of a particular kind of society. It is not as neutral as it might appear.

The objectivity which seems to be guaranteed by a God's-eye view itself suggests a form of detachment. One of the prized virtues of science is its objectivity, its apparent willingness to be led by evidence alone and not by prejudice. Yet even here absolute detachment from all points of view is impossible. The breakdown of positivism and the emphasis on the priority of theories has only served to emphasize how a scientist who is totally detached from any conceptual scheme will have no means of discriminating between parts of reality. Science without concepts is blind. The objectivity it has to aim at is not the detachment of the umpire, who supports neither side, but the objectivity of the searcher after truth, who is only concerned to discover what is in fact the case. There is no virtue in being detached from a true theory. The problem is knowing which one it is. The desire for a neutral point beyond all theories is mistaken. It can only serve to foster the idea that objective truth is impossible to come by. If truth in science is a property of theory, and objectivity were to demand detachment from theory, we arrive at the absurd conclusion that objectivity and truth, so far from being virtually synonymous, are in fact opposed to each other.

Yet if there is no standpoint beyond theory, we may seem to be trapped by the presuppositions of whichever one we are committed to, however imperfect they may be. Where does that leave the question of objective reality? If reality is what a theory says it is, the change from, say, classical to quantum mechanics involves a change in reality. We seem then to be living in a different world, with all the problems of incommensurability arising. What is needed is the idea of something independent of the scheme. The problem seems to be that if there are only partial viewpoints available, with no universal viewpoint, exemplified by the image of a God's-eye view, there seems no way of transcending the limitations of one's particular context.

A parallel can be drawn in the controversial field of theories of justice. One response to the vexed problem of contested views of the nature of justice is to accept that there are different ideas of what is just. It is, however, very tempting to conclude that somehow it is just not to interfere with the varying interpretations. Michael Walzer is ready to talk of different worlds in this context, and says explicitly that 'justice is relative to social meaning'. He continues:

> We are (all of us) culture producing creatures: we make and inhabit meaningful words. Since there is no way to rank and order these worlds with regard to their understanding of social goods, we do justice to actual men and women by respecting their particular creations. . . . To override these understandings is (always) to act unjustly. (1983, p. 312)

Yet the claim that justice is rooted in shared social meanings does not sit very easily with the idea that overriding such understandings itself involves justice. Either what is just is relative to a society, or there are objective criteria for justice that transcend societies, and guide us in how to treat varying communities. It does not seem possible to hold both conceptions simultaneously. Our inability to rank different perceptions of justice may suggest the impossibility of a God's-eye view, but Walzer's warning against overriding different understandings seems to be an assertion of something as true. This neatly encapsulates the difficulty of anyone confronting the temptations of relativism. We cannot get outside all conceptual schemes, and yet the very project of talking loftily of varying conceptions of justice implies an ability to abstract ourselves from all possible contexts. Saying that every community is as good as every other implies that we can stand above or apart from all particular communities and talk about them. The espousal of a relativist position itself implicitly trades on some idea of a God's-eye view. It does this not in order to see which of various rival positions is correct, but to survey them all at once with apparent neutrality. In fact, the ability to abstract ourselves from all possible claims to truth is a dubious advantage, particularly if we still wish to make assertions about what is the case. There seems no way of avoiding a deep incoherence. We have stood apart from claims to truth only, it seems, to make some other claims. Detachment from all schemes, even if it were possible, would inevitably result in judgements being made about them, even if only that we should not interfere with them. We cannot opt out of all judgements about what is the case. The idea that there can be any position of perfect neutrality outside all schemes is totally illusory. Adopting a liberal approach is itself to hold a substantive position, making a definite judgement about what is important. It gives more weight to individual freedom and equality than to particular conceptions of what makes a good society.

A God's-eye view is not in fact about neutrality at all. Those who invoke it do not just seek detachment. They want perfect knowledge. The idea implies that the most important question is how we, or at least someone, could ever know what is true, or good, or just. The question is how we could attain complete knowledge, and since the task seems beyond us, we seem to be caught in our own particular context. It seems as if because we cannot be like God (if indeed there is a God), we cannot aspire to transcending the particular historical circumstances in which we are placed. Yet the possibility of knowledge, particularly of human knowledge, is a question for epistemology. Appeals to the apparent necessity of a God's-eye view can be a subtle way of reducing metaphysics to epistemology. It changes the subject from what is real to what can be known, and even to whether we can know it.

Reference to God in these contexts is doubtless not intended
to be particularly theological. It is meant to invoke the possibility
of omniscience, or of detachment from the confining perspective of particular contexts. God just becomes an ideal observer,
disinterestedly taking note of everything. Nevertheless, this can
be a source of confusion, since *if* there is a God, He is the ultimate cause of whatever exists. By definition, He is the source of
all meaning, and the ultimate explanation of reality. Yet the phrase
'a God's-eye view' does not imply any of this, since its emphasis
is on knowledge. It is concerned with what God knows and not
His sustaining of reality. How far we can attain knowledge is a
very different question from what would ground our reason and
give it a context beyond the immediate influences on us. Mention of God could give us confidence that reasoning towards the
truth is a possible human activity. That is what Descartes thought,
both because reality was, he thought, grounded in God, and
because we could aspire to reflect the rationality of God. If He
has given us reason, and created a universe which is rationally
ordered, we ought to be able to understand something of the
world around us. We are made in the image of God and so can
begin to discover some of the structure and order inherent in
His creation. This is a doctrine, though, about reality and our
place in it. It is in no way simply a view about the possibility of
knowledge. The latter is explained in terms of nature of reality.
As always, epistemology has to depend on a prior metaphysics.

The Search for an Ultimate Ground

Theology tries to give an ultimate ground for reality, and an
explanation why we have access to its nature through reason.
The hope is that we then have an assurance both that there is
something to be known, and a way of coming to know it through
the exercise of our God-given reason. That reason might be
expressed in the methods of science, but, could also itself go
beyond the limits of what is empirically given. This is a very
traditional way of looking at human rationality and one that sets

out to justify the possibility of philosophy. The more recent tendency has been to keep assumptions about the possibility of rationality and the grounding of science, whilst repudiating the theological background which gave them plausibility. All this is now being challenged, in particular by 'post-modern' views of philosophy. We have to examine again what makes human rationality possible. What enables us to abstract ourselves from the historical conditions which have helped to mould us so that we can indulge in what Rorty (1991b, p. 52) patronizingly calls 'arm-chair philosophy'? He sketches how an atemporal deity was replaced in Kant's philosophy by the idea of an atemporal subject of experience. We could still, it seemed, rise above our particular historical circumstances and indulge in 'pure' reasoning. Later on, he suggests, 'language' replaced 'experience' and 'mind', but even after the so-called linguistic turn, when human language seemed to encapsulate human rationality, philosophy still seemed to have the resources to transcend local limitations. According to Rorty, Kant's study of the conditions of experience was replaced by an emphasis on the conditions of describability in language. Philosophy seemed to be able to deal with such conditions, and to talk about all possible languages. This, however, itself presupposed an ability to abstract from the local conditions of an actual language. It suggested that one could obtain that elusive God's-eye view, even if this time we were surveying languages rather than the world.

It is a mistake, however, to assume that in order to talk of the conditions of possibility we have to have a panoramic view of all language. All we need to do is to recognize that reality, whatever it may be like, must act as a constraint on all language. Language, and the reason it exhibits, must be grounded in the way things are. We need an explanation that will not immediately generate the need for another. An explanation that starts off an infinite regress cannot be satisfactory. Rorty distinguishes between different types of entities referred to in the history of philosophy. One type 'requires contextualization and explanation but cannot themselves contextualize or explain' (1991b, p. 54). Examples could be intuitions for Kant or material objects for Plato.

The former could be made knowable in some way by the Kantian categories, and the latter by the Platonic Forms. The point of this is that the categories and Forms are intended to explain without themselves needing explanation. In the case of the Forms, for instance, Plato explained the nature of material entities in terms of the way they 'shared' in the Forms. Something was good because it shared in Goodness. What though of Goodness itself? Plato himself saw that if Goodness was itself good (and as an ideal standard it would have to be), it would then have to share in a third entity which covered both the Form and its particulars. This 'third man' argument seemed to generate an infinite regress, so that the source of explanation had to be explained. If the Forms have a magic ingredient which makes them invulnerable to a need for further explanation, there is no reason why individual instances should not have it too.

The same dilemma occurs in theology, and Rorty asks why the world cannot be its own cause, if God can be thought to be His own explanation. Once we have accepted the possibility of unexplained explainers, anything can be seen to be in the same situation as a transcendent deity. We can believe in things without, he says, 'relating them to something which conditions their existence or knowability or describability'. If some things can be accepted without our searching for reasons why they are available, why should not others? The pursuit of justification no longer seems so essential. Yet as Rorty himself sees, this is in effect to give up on the project of explanation and justification. He says that 'we have thereby questioned the need for philosophy, insofar as philosophy is thought of as the study of conditions of availability' (1991b, p. 55). It is perhaps significant that Rorty talks of 'availability' and runs together questions of existence, knowability and describability. Indeed, although there may be structural similarities between Kant's treatment of rationality and Plato's doctrine of Forms, the latter is intended to be firmly grounded in the way things are. Kant is dealing with the conditions necessary for the possibility of human reasoning. There is a great deal of difference between questions about existence and issues about how far we can know and describe things. Language has to be grounded in something beyond itself, and we cannot know without knowing something. Existence is in one sense its own explanation. That something exists can be accepted as a brute fact. Even if the meaning and purpose behind existence is attributed to God, arguments for the existence of God can start by taking the existence of some things at face value. We do not need to know whether God exists before we accept the fact of the existence of the world around us.

Purported knowledge and attempted description of things require justification. A connection has to be established between a claim to knowledge, or the description of reality, and the way things are. Yet such justifications will merely generate an infinite regress if we justify something in terms that themselves require justification. Existence is not like that. We do not need to rest a claim that something exists on a claim that something else does. We may well wish to explain why something exists, or how it came into existence. A motor car is not its own explanation, but was manufactured for a purpose. Nevertheless if a car is outside in the road, that fact can be accepted as such without our having to know why it is there. Yet a claim to knowledge is something that does require justification, not least by an appeal to the way things are. We must be correct in our estimation of reality, and that correctness must not be just a lucky accident but spring in some way from the nature of reality.

It might be claimed that a dichotomy between knowledge and existence is totally misguided. There may seem in the end to be no difference, if we cannot talk of what exists without claiming knowledge. We know of no other method of access to existence. Even if the fact of existence does not require grounds, our knowledge of that fact does. We cannot grasp the fact without knowledge. Yet this is to lose sight of the point that our knowledge is grounded in reality. That this may seem puzzling springs from the anthropocentric character of so much modern philosophy, which never allows us to forget that when we talk about reality we are only using *our* concept of reality. Yet once this has been accepted, we can no longer think of the way our concepts and beliefs might be grounded. We are just caught up in the web of our beliefs. Philosophy will be powerless to provide any external justification for our reasoning. As Rorty's work suggests, this can easily spell the end for philosophy as it has been traditionally understood.

Internal Realism

The realist emphasis on reality as the focus and grounding of human rationality has fallen into disfavour. Hilary Putnam warns of the vacuity of what he terms 'metaphysical realism'. Instead he advocates an 'internal realism', according to which reason, although understood as immanent within a particular conceptual scheme, can also claim a transcendent validity. Putnam stresses the fact of conceptual relativity, the way in which all our claims are shot through with the assumptions of a particular conceptual scheme. As a result, it seems inevitable that we can only characterize what is real through the terms of one scheme or another. He goes so far as to take it as a fact of life 'that the enterprises of providing a *foundation* for Being and knowledge – a successful description of the Furniture of the World or a successful description of the Canons of Justification – are enterprises that have disastrously failed' (1990, p. 19). It is significant that Putnam regards Being as itself needing a foundation, since it should surely *be* the foundation. Yet Putnam is influenced, like others, by American pragmatism and will only look at reality from the human point of view. He could never ground our beliefs in reality, since he would want to ask how we can conceive of that reality and describe it. Reality, to be meaningful to us, must, in his view involve successful descriptions of it. This comes over explicitly in the characterization of the 'metaphysical realist'. Putnam writes: 'What the metaphysical realist holds is that we can think and talk about things as they are, independently of our minds, and that we can do this by virtue of a "correspondence" relation between the terms in our language and some sorts of mind-independent entities' (1983, p. 205).

This definition of realism immediately induces a reference to 'us', whoever 'we' are. It paves the way for a withdrawal from talking of us as rational beings to a view of us as individuals placed in a particular historical situation. There has to be something radically wrong with a metaphysics that claims to be about reality but turns out to be about how we can engage with it. Philosophers can easily show the inherent tension involved in our thinking of what is independent of our minds. The presumption that we can contact reality lies at the heart of all thought and language. Unless, however, there is a guarantee that our concepts can reflect the way things are, it is plausible to suggest that such concepts are the product of the context in which we are placed. Yet metaphysical realism, if it is to be worthy of its name, must be a doctrine about reality and not our contact with it. Least of all can it be a theory about our language. There are philosophers who only view metaphysics through the functioning of language. To them it is self-evident that arguments about realism involve us in the examination of our linguistic practices. Yet once this step is taken, we inevitably have to conceive of reality from the speaker's or agent's point of view.

Michael Dummett, for instance, believes that the true nature of metaphysical disputes about realism is that 'they are disputes about the kind of meaning to be attached to various types of sentences' (1991, p. 14). Yet sentences in a language are sentences used by people. Questions about the states of reality thus are turned into questions about people's understanding. We have stepped back from reality to the way it is spoken of. All philosophy depends on the medium of language, but it is still a great mistake to imagine that metaphysical problems are in fact problems about language. Language is a human construction. What we try to talk about most assuredly is not. There are always problems about how our reason, which is normally encapsulated in language, can deal with what is the case. The more, however, that we become preoccupied with language which should only be our tool, the more we will lose any grasp on the reality it exists in part to describe. Dummett insists that a picture of reality has no content beyond the model of meaning which it suggests. He continues: 'However powerfully the picture impresses itself on us, we have to bear in mind that its content is a thesis in the theory of meaning, and that, beyond that, it is no more than a picture' (1991, p. 15). Dummett's view is that models of the meaning of sentences belonging to different sectors of language will elucidate how those sentences are to be counted as true. It will thus 'adjudicate between the rival conceptions of truth advocated by realists and anti-realists' (1991, p. 14). The issue, though, must always be not what we mean in our language, but what is the character of the world which we encounter in our use of language. The two problems are not the same, as is made apparent by the feeling we often have of the inadequacy of language to deal with what confronts us. Concentration on meaning can easily be accompanied by the tacit assumption that our world is constructed by language. The realist will always deny that, and yet making realism a matter internal to language can prevent this basic issue being articulated.

The guiding principle behind Putnam's 'internal realism' is that, in a sense, we view reality from the bottom up. We arrive at it through our practices, and do not start with it as something clearly there. Truth is a linguistic category, and becomes a projection of the standards of rational acceptability which are established in our language. Putnam maintains that metaphysical realism champions the idea that the world consists of a 'fixed totality of mind-independent objects' (1990, p. 30). However, this is not a sufficient characterization for him. He adds two other marks of that kind of realism, namely that there is exactly one true and complete description of the way the world is, and that truth involves some sort of correspondence. Questions of being are quickly transformed into questions of language. The later Wittgenstein's view of meaning as use can easily be combined with the pragmatist view of truth as warranted assertibility, and Putnam owes much to both sources. He writes: 'Access to the world is *through* our discourse and the role that the discourse plays in our lives; we compare our discourse with the world as it is presented to us or constructed for us by discourse itself, making in the process new worlds out of old ones' (p. 121).

The minute that a philosopher refers to the construction of new worlds with linguistic tools, any fully-fledged doctrine of realism may seem to have been left behind. This is certainly the case in the physical sciences. Matters are more complicated in the social sciences where society must itself be regarded as a human construction. Even there, however, social theory will not create the social reality with which it is concerned. Putnam seems to have given up the idea that there is one world waiting to be discovered. Instead, many different worlds can be constructed. The whole possibility of metaphysics has been removed. Although it is conceivable that there are many actual worlds, and in a sense even many physical universes, if there are, they are not constructed by us or our language. It would seem more sensible still to talk of one world with different isolated regions or phases. Even possible worlds are not simply the product of our imagination. Their possibility should arise from the nature of reality and not from our ability to conceive them.

Putnam's internal realism, with its alternative conceptual schemes, becomes very removed from the traditional realist view. This can be seen from his admission that in his picture 'objects are theory-dependent in the sense that theories with incompatible ontologies can both be right' (1990, p. 40). He refers to James and Dewey and says that languages and theories, functionally equivalent in their context of inquiry, 'are equivalent in every way that we have a "handle on"'. In true pragmatist fashion, truth becomes 'idealized rational acceptability', and the issue is simply one of what criteria we have to hand. Metaphysics is not a possible subject for Putnam (p. 39). This is because of his conviction that any metaphysical system 'will have to be accompanied by some sketched-out story of how we can have access to "metaphysical reality"'. This is, of course, the perennial problem. Metaphysics may have pretensions to be a pure science of being, but it must itself be a human view. Yet, once we begin to worry about our access to reality, we are concerned with human capabilities and not with reality. We need a concept of what may lie beyond our concepts, but this is still one of our concepts.

The idea of a God's-eye view might provide the needed contrast with our limited conceptual scheme. Putnam, however, has been as responsible as anyone for the repudiation of this approach. Moreover, he has explicitly associated this with the impossibility of metaphysical realism. He says: ' "The God's-eye view" – the view from which absolutely all languages are equally part of the totality being scrutinized – is forever inaccessible' (1990, p. 17). He is particularly critical of the ideal of impersonal knowledge which it apparently champions. It seems to suggest that there is 'a point from which we can survey observers as if they were not *ourselves*, surveying them as if we were, so to speak, *outside our own skins*'. It considers that we can never conceive of things as they are apart from us. Our ways of thinking are indissolubly linked with how we describe the world. We can claim truth, but our claims are tethered to their origins. In that sense, Putnam is not a relativist, because he does not wish to limit the scope of our claims to truth. Indeed he attacks relativists for thinking that they can stand outside their language, in order to talk of absolute limitations. He says of Rorty that 'the attempt to say that *from a God's-eye view there is no God's-eye view* is still there, under the wrapping' (1990, p. 125). Yet in the end, Putnam's repudiation of metaphysics is as complete as Rorty's. Pragmatism wins the day.

Putnam in fact claims 'to see relativism *and* the desire for a metaphysical foundation as manifestations of the same disease' (1992, p. 177). One should not feel the need for such foundations, nor even deny the need, as the relativist does. Instead, Putnam thinks, we should be willing to go on using our language-game. We should not be afraid to say that some things are true, or warranted or reasonable. He believes that 'we do have the language, and we can and do say so, even though the language does not rest on any metaphysical guarantee like Reason'. In fact, following the later Wittgenstein, Putnam says that 'our language game rests not on proof or on Reason, but *trust*'. We must, it seems, accept the world and other people, 'without the guarantees'. We must go on with our practices and language-games, and resist the temptation to stand outside them, and make a comment about them. Such a project is impossible, Putnam believes, since talking *about* our language-game is already to distance ourselves from it.

In all this, the pragmatism of philosophers such as Dewey is explicitly combined with the Wittgensteinian absorption in language-games and forms of life. The common thread is the refusal to allow reason to become detached enough from our particular context so that we can reason *about* it instead of merely from within it. Internal realism in effect demands the impoverishment of reason. 'Trust' is all very well, but it becomes very hard to distinguish it from a complacent acceptance of whatever conventions hold sway where we are. The repudiation of the possibility of a God's-eye view can paradoxically also involve neglecting the power of reason. We become trapped within the bounds of a practice, system, language-game, or whatever. According to Putnam, 'the great contribution of Dewey was to insist that we neither have nor require a "theory of everything"' (1992, p. 187). In this context, that means something like a God's-eye view or an absolute conception. Yet even without it, there should still be room for the possibility of rising above our immediate circumstances. We should not be forced into a choice between reasoning about everything and reasoning about nothing. Just because we cannot possess omniscience, does that mean that the only alternative is to be content with the prejudices of our time and place? The function of metaphysics is surely to warn us of the dangers of such parochialism. Where we are is not necessarily where we ought to be. We can reason about our circumstances and our position in them, even if we can never leave them completely behind.

Realism and Physicalism

One of Putnam's motives for denying a role for metaphysics is to give a place for 'values'. Scientific emphasis on so-called 'facts' has made 'values' look very subjective by comparison. Philosophy is still haunted by the crude division between what can be known and what appears to be merely the object of individual preference. That is the legacy of positivism, and it is to Putnam's credit that he wants to reinstate ethics, so that it is no longer the poor relation of science. His strategy, however, is in effect to downgrade science rather than upgrade ethics. He wishes to cast in doubt the particular connection with the real world devised by science. Instead, we are to concentrate on our actual practices, whether scientific, moral or whatever. He does not believe, he says, that 'the metaphysical realist picture has any content today when it is divorced from physicalism' (1990, p. 37). Given this view, it is inevitable that, once one is convinced that physicalism ignores too many important features of human experience, one will jettison realism as well. Yet physicalism does provide a systematic account of the world, and our place in it as physical objects related to the rest of the physical world. It is perhaps hardly surprising that Putnam should regard it as the only genuine candidate for a substantive realist position.

We saw in the last chapter how physicalism deals exclusively with the entities and processes laid down in physical theory. As such, it is essentially a commitment to the findings of science, particularly physics. Yet without the assumption of a reality to be investigated, the practice of physics should not even begin. That assumption is not the product of enquiry into the nature of the physical world, but is a precondition for it. Even if the character of reality is such that it can be wholly expressed in physical terms, this goes beyond the scope of actual and even possible human investigation. It is itself a metaphysical principle which would seem to call for rational support. Putnam, in fact appears to think that physicalism is actually mistaken, because it undermines many human practices. Instead of accepting that he is himself indulging in metaphysics, he prefers to attack the possibility of metaphysics on the grounds that that will involve us in physicalism. This is all the more curious since the latter is normally considered the enemy of metaphysics.

Putnam attacks 'objectivism' which he sees as the identification of reality with the discoveries of science. He considers that there are two basic 'objectivist' assumptions (1987, p. 15). The first is that there is a clear distinction to be drawn between the properties things have in themselves and those projected by us. The quest for such objectivity has haunted philosophy, and has been a major cause for dispute in science itself, as in the debate over quantum mechanics. It is the particular concern of metaphysical realism. Putnam, however, adds a second assumption to the effect that 'the fundamental science – in the singular, since only physics has that status today – tells us what properties things have "in themselves"'. He derides this objectivist picture as one of 'nature as the World Machine'. That would involve, though, a rapid slide from a conception of an independent reality, to one of it as exclusively physical, and from there to a view of it as operating according to fixed and rigid physical law. Not all reality may be physical, and physical reality may not be at all machine-like. It is important, too, to resist any easy assumption that the concept of determinate reality with an inherent structure has anything to do with any doctrine of determinism.

Putnam considers that 'science is wonderful at destroying metaphysical answers, but incapable of providing substitute ones' (1987, p. 29). His conclusion is that 'science has put us in the position of having to live without foundations'. Yet science needs foundations. This does not mean we have to be foundationalists in epistemology looking perhaps for incorrigible experience as the basis of our theories. The foundations needed by science must be provided by reality and not by our experience of our concepts. Many will, as always, resist this separation of reality from understanding. This is the thrust behind Putnam's internal realism. The way things are in themselves can only be the way they are judged as being from the point of view we take. This, though, is too tolerant of conceptual relativity. When people diverge, they cannot all be right. Realism insists that the world has a determinate nature. The idea that objects must in any way be thought relative to our discourse or conceptual scheme certainly involves the denial of the possibility of metaphysics, except perhaps in so far as it grows out of our present linguistic practices. Putnam's denial of the possibility of a God's-eye view leads him to the opposite extreme of rooting us too firmly in our present circumstances. Yet human reason can transcend its limitations, and Putnam occasionally accepts this. He is opposed to relativism and says: 'We always speak the language of a time and place: but the rightness and wrongness of what we say is not *just* for a time and place' (1987, p. 21). Yet because he rejects the idea of a reality common to all competing conceptual schemes, he is forced to talk in terms of Dewey's phrase, 'warranted assertibility', rather than of justification. He looks to our actual practices, which have evolved through time, to provide the setting for the claims we make.

'Things' have always, it seems, to be mediated through our concepts, and this means, according to Putnam, that we are immediately referred to the character of the practices which have produced the concepts. In particular, he says 'our norms and standards of warranted assertibility are historical products' (1990, p. 21). Although he accepts that there are better and worse norms and standards, our attention is remorselessly drawn away from how these could be distinguished to the particular facts of history, and to the circumstances of our specific practices. Yet even the practice of history demands an ability to detach oneself from the prejudices of one's age and to transcend one's limitations. That is part of what is meant by having a God's-eye view, since transcendence is a key property of a deity. Yet the possibility of any proper reasoning demands it, not as an end in itself, but as a means of putting aside the distractions of irrelevant prejudice.

Like many people, Putnam takes science to be the test case for realism. He says: 'If science does not tell us what is "really there" in the metaphysical sense, then neither does anything else' (1990, p. 90). The trouble is that this attitude can so easily encourage the identification of realism with physicalism. Yet it remains true that scientific methodology seems to equip us particularly well to uncover the workings of at least some aspects of the world. How can this be explained? Realism can certainly talk of scientific success in a way which is impossible for more anthropocentric positions. We are able to predict and control parts of the world because our theories have put us in touch with the way things actually work. Certainly, the transcending of our historical situation and an ability at least to glimpse aspects of a world that is not merely the projection of our prejudices would seem an inescapable requirement for the practice of science.

A God's-eye view as a means of seeing things as they are, is a worthy ambition for scientists. It is, however, not a part of a definition of reality, but only a striking image of the need to break free from the limitations inherent in the human condition. It is all too easy to cry down the possibility even of attempting to do so, but if we remain content with our present conceptual scheme, the result will be a stultifying complacency and paralysis. Nowhere is this more obviously true than in the conduct of physical science. We may well know some things now, and new knowledge may be within our grasp. We should not follow Kant in imagining that all things in themselves are for ever beyond our reach. An unfortunate connotation of the notion of a God's-eye view is that such a state is by definition unattainable here and now. Although we will never know everything, it does not follow that we cannot come to know some things.

Putnam's own view about the success of science is that it has to be a puzzle, as long as concepts and objects are viewed as radically independent of each other. In other words, science must be made successful by definition, by making a logical connection between concepts and the reality they describe. It seems as if to some extent, 'objects' are projections of the concepts. The two are not truly distinct from each other. Putnam warns us of thinking of the world 'as an entity that has a fixed nature, determined once and for all, independently of our framework of concepts' (1990, p. 162). It is all too apparent, though, that on this issue depends the very possibility of metaphysics, of grounding science in anything beyond itself, and of explaining why 'success' in science is a genuine achievement. The idea, however, that the world is determined once and for all seems to carry wide implications. It is perhaps no wonder that Putnam associates a God's-eye view with a materialist metaphysics which sees the universe as a closed system. That is totally different from a realist conception of a determinate world, which has an inherent structure and order. We may do well to be suspicious of the old mechanistic and deterministic view of the world. The advances of contemporary science certainly suggest as much. This, though, is not a matter of the conceptual 'net' we place across the world, but should be because of the actual character of reality itself. Otherwise, there is little point in changing the net. The inherent nature of the world is such that earlier materialist views can now be shown to be inadequate. They exaggerated the importance of some regularities and ignored other aspects. Good science, however, can only be grounded in the way things actually behave and not the way we assume they do. Reality can and eventually does have the last word.

6
Science and Humanity

Concepts and Causes

Whenever the objectivity of reality is stressed, the problem always arises as to how we humans can get in touch with it. An emphasis on the importance of a God's-eye view may seem to remove reality forever from our grasp. Even if this is seen to be irrelevant, we are still going to be faced with the question of how we can now understand what exists. How can our minds gain knowledge of an existence which is totally independent of them? Idealists try to tie reality to our understanding by definition. In this chapter we shall also see how some scientists use theoretical physics to connect the very existence of the universe with our existence in it. It is a project which, at times, itself risks becoming a species of philosophical idealism.

Science is a human product. It is the result of human reasoning and would not exist without human minds. This may seem a truism, but if emphasized too much, scientific theory will seem to reflect facts about us rather than about the world. The stress on science as a human form of life shows what can happen if we take our eyes off the basic conundrum of the relationship between ourselves and the rest of the universe. In part, this has been what the argument between idealists and realists has always been about. The idealist view of the primacy of mind can easily make reality appear to be the reflection of the human mind. At best, reality recedes beyond our grasp, like Kant's *noumena*. It may exist, but can never be conceived as it is. There are many who are reluctant to proclaim the unreality of the physical world, and yet who are impressed by the active role the mind seems to have in making sense of its surroundings. For them a form of conceptual idealism can appear attractive according to which our descriptions of the world are necessarily mind-involving. Rescher, for instance, points out that even if there could be a physical object indistinguishable from a book, 'there could not be *books* in a world where minds had never been in existence' (1991, p. 497). He goes on to insist that the world for us must be as we conceive it to be, and that even such basic concepts, as that of causation, are implicitly mind-referring. By definition, he says, a 'mind-untouched' reality is one 'about whose nature we neither do obtain nor can obtain any knowledge'. Yet books are one thing, and causes quite another.

A realist could cheerfully acknowledge that books are what they are because of human minds. They have been produced by minds, in the quite straightforward sense that their contents are the result of human thought and can only be understood through mental processes. Yet once written, they do have an independent existence. They are not dependent on minds in any logical sense, as the idealist would have it, but in a purely causal way. We would not even be able to understand what a book was, unless we could recognize the two-way relationship between minds and what is written on the page. It is highly misleading to suggest that a cause is in any way similar. Causes concern the relationship between separate events in the world. An idealist may certainly allege that we do not interpret what *really* happens but bring our own conceptual scheme to bear upon what would otherwise be a formless chaos. Causation would then be a feature of our association of different events rather than a reflection of real connections between real objects in the world. Indeed the whole point of different forms of idealism is to resist the possibility of any such contact. Rescher summarizes a favourite argument in this connection: 'There is no clear and sharp separation between reality and the domain of thought, because our only possible route to cognitive contact with "the real world" is through the mediation of our conceptions about it, so that, for us, "the world" is inevitably "the world as we conceive it to be"' (1991, p. 512).

This approach characteristically attempts to build a major philosophical theory about our relationship with the physical world out of the blindingly obvious truism that we cannot think of what we cannot think of. Put more positively, the view is that we can only think of reality in the way that we can think about it. Yet although our concepts are certainly the means through which we approach the world, we are not talking about *them*, when, with their aid, we try to refer to parts of reality. The concept of a cause may be a product of the human mind, but unlike books, causes can operate totally independently of minds. Indeed it is an unfortunate fact of human life that we are often profoundly ignorant of what is influencing our immediate environment. Empiricists such as Hume, show their idealist leanings when they are unable to give a satisfactory account of causation, other than in terms of our own expectations.

Rescher realizes his theory of conceptual idealism runs into problems when he tries to describe causal interactions in the world, particularly between matter and mind. He distinguishes between moving from mind to matter 'in the conceptual order of understanding' and from matter to mind 'in the explanatory order of causation' (1991, p. 514). The problem, though, is whether the two can be asserted sumultaneously. Can an idealist talk of the causal effects of matter on mind while admitting that the very idea of causation is a mental construction? Without great care, the whole notion of cause can quickly disappear. Instead of helping us to account for what happens in the physical world, it becomes understood merely as a way of organizing our experience of the world. If we wish to invoke the idea of cause to explain the latter, as being the result of causal processes, we will be involved in a circle which we cannot break out of. This certainly may appear to trade on precisely the realist concept of cause which idealists reject. Nevertheless, Rescher, for one, wants to draw a distinction between our conceptual understanding and causal explanation. He holds that conceptual idealism is an analytical theory about our categories of understanding, not a theory about the causal mechanisms underlying the mind (1991, p. 515).

The problem is whether the two are compatible. Can one have a proper idea of causation, which perhaps could even involve explaining the origins of mind and its connection with the physical world, and still make the whole idea a mere mental construct? Causation is hardly a genuine explanatory notion if it does not refer to real processes. If it is just one of a network of concepts, we cannot give any causal explanation of our conceptual scheme. On the other hand, if we can break out of our conceptual prison and talk of the real world when we refer to causal relationships, doubtless we can do the same with other concepts.

A strength of conceptual idealism, not to mention stronger forms of idealism, is that it gives us a ready-made answer to the perennial question as to why we can understand the world. Because the world, which we describe, is *our* world, delineated by our concepts, there can be no mystery why our understanding matches it so well. There is a sense in which it has been created in our own image. Once however, we turn to the physical world as something existing in its own right, the problem of the relationship between our understanding and what we are trying to understand becomes very great. We have to face the question of our place in the scheme of things. Because idealists start with the mind, and that generally means the human mind, their views are unacceptably anthropocentric. The problem for realists is that we have to face our apparent insignificance given the immensity of the universe. There seems no reason why we should expect to be able to understand its inner workings. It is even possible that we cannot know anything. Indeed it was to prevent such sceptical reflections that pragmatists insisted that we start from where we are, and empiricists insisted on the authority of human experience. We have also seen how a naturalistic epistemology has attempted to give some reason for a realist to trust concepts that have stood the test of time, by appealing to evolution.

The Unity of the Universe

Philosophical theories, such as idealism and realism, can often seem to be mere accounts of the meaning of language to us.

They are hardly strategies which measure up to the grand task of relating human beings to the cosmos. It may be that the very idea of locating humanity in such a way is incredibly arrogant. Have we not, it might be said, long grown out of the pitiful notion that the universe was made for us, and that human beings are the centre of creation? Our planet is insignificant, orbiting a minor star in one galaxy of one hundred billion stars, in a universe of countless galaxies. If we take contemporary physical theory seriously, as no mere reflection of the human mind, but as an attempt to describe and explain the very constitution of the cosmos, it seems to reinforce the view that we count for nothing. Yet we must face the paradox that apparently our small reason can come to grips with the immensities of space and time. The claim is that somehow *we* can actually understand some of the workings of the universe. The light of human reason can, it seems, illuminate the further regions of space–time up to the very origins of the universe. Yet if this cosmos is not simply the creation or reflection of the human mind, how can the human mind come to grips with it? The Greek word 'cosmos' implies an ordering, in the way, say, that people might adorn themselves. (The word 'cosmetic' comes from the same source.) The question, though, is why anyone should imagine the universe is ordered, or at least behaves in an orderly and regular way. The very word 'cosmology' already builds in this assumption. The comprehensibility of the universe cannot be taken for granted, and yet the very existence of science as a discipline depends on the fact that it is comprehensible. This not only means that it is enclosed and structured, but that it is ordered in a way that is intelligible to us. One physicist has written: 'Through conscious beings the universe has generated self-awareness. This can be no trivial detail, no minor by-product of mindless, purposeless forces. We are truly meant to be here' (Davies, 1992, p. 232).

Why though should the fact of human awareness and understanding be of such importance? We return to the question of the relations between the human mind and the universe of which it is such a small part. A possible conclusion to draw is that whatever consciousness and rationality amount to, they are features which have evolved by chance, and cannot possibly be related to important questions about the nature of the universe. Yet this actually contradicts theories about the nature of the Universe which starts with the Big Bang model of an expanding universe. It is generally accepted that life has needed time to develop. Our bodies, for example, need carbon, in a complex relationship with elements such as nitrogen, phosphorous and oxygen. These were produced in stars, which in their death throes as supernovae scattered the elements through space. In turn, they were incorporated into planets and ultimately serve to constitute our bodies. As John Barrow puts it 'we are the ashes of the stars' (1988, p. 354). The catch is that stars explode only after burning their nuclear fuel, and this takes around ten billion years. The size of an expanding universe is a function of how long it has existed. There would be no carbon to help provide the building blocks of life in a universe that was much smaller than the present one and had therefore existed for a shorter time. Barrow says:

> A Universe that contained just one galaxy like our own Milky Way . . . could have expanded for little more than a few months. It would have produced neither stars nor biological elements. It could contain no astronomers. . . . The universe needs to be as big as it is to support just one solitary outpost of life. It is a sobering thought that the global and possibly infinite structure of the universe is so linked to the conditions necessary for the evolution of life on a planet like Earth. (1988, p. 354)

When this line of reasoning is pursued, it becomes clear that for the universe to be as it is and support life on Earth, its structure had to be fine-tuned to an exceptional extent from the earliest moments after the Big Bang. If any physical constants were only slightly different, the universe would not have expanded as it did, creating stars and planets and the chemical elements necessary for our own existence. Modern physical theory shows how life on one planet is inextricably bound up with the basic structure of physical reality. Physicists can point to a string of cosmological coincidences, relating to the basic physical forces which join in providing the physical conditions necessary for life. For example, if the density of matter were greater than it is, the expanding universe would collapse. If it were smaller, expansion would be too rapid for planetary systems to maintain any stability. The conclusion has to be that human life is not simply an unaccountable accident unrelated to the basic physical processes of the whole universe. Our existence depends on what happened even in the moments immediately after the Big Bang.

John Wheeler, one of the most influential figures in modern cosmology, sums up the situation as follows:

> It is not only that man is adapted to the universe. The universe is adapted to man. Imagine a universe in which one or another of the fundamental dimensionless constants of physics is altered by a few percent one way or the other. Man could never come into being in such a universe. That is the central point of the anthropic principle. According to this principle, a life-giving factor lies at the centre of the whole machinery and design of the world. (Barrow and Tipler, 1986, p. vii)

It is easy to rush into talk of 'design', but we should be cautious about what precisely is shown by the anthropic principle. Some physicists are apt to claim that the Universe knew we were coming, and that everything was geared from the very beginning to the production of life, even human life. Others may point out that we would hardly be able to describe a universe which was inhospitable to life. We would not be here to describe it. There can easily appear to be a sense of things being explained after the event, of a purpose being seen in everything that just chanced to happen. It is sometimes assumed that we are meant to be here because we are here. Yet we could hardly live in a universe that was not able to produce life. Why, too, is it so readily assumed that we must have a privileged position in the constitution of the universe? We undoubtedly think we are important, but it does not follow that we are. Why, in other words, is the principle an *anthropic* one, rather than, say, merely a star principle or a carbon one?

These may not be unanswerable questions, but we should be wary of assuming that because we exist, everything was engineered to make our existence possible. Yet no theory of the universe that rules out the possibility of human life is tenable. A general characteristic of anthropic arguments is that they ask what kind of a world it must have been to have produced people like us. Since we are self-conscious and able to produce theories of the universe, it is natural to link the existence of such rationality to the pre-conditions of any universe. We could not produce theories about a universe that could not produce rational thinkers. Arguments expounding variations of an anthropic principle are linked to issues concerning self-reference. No theory that undermines itself can be supported, and so we cannot produce a physical theory that cannot account for our own existence. One feature of anthropic arguments is that they demonstrate that the existence of life is not a minor issue, which is irrelevant to physicists and can be left to biologists, since its constituents are intimately related to the fundamental nature of the universe. The universe we inhabit is an impressive unity.

Many Worlds

Any physical theory must start with the acknowledged fact that we are here and able to reflect on the nature of the universe. Our universe has to have been able to produce life. That, however, does not prove that there could not have been, or indeed may not be, other universes with very different properties. We may accept that there is a link between the fact of our observation and the kind of universe we live in, but all that may follow is that some universes would be such as not to produce observers. Barrow says that 'this recognition, that there are types of universe which we could not expect to observe, is often called the *Weak Anthropic Principle*' (1988, p. 355). This is in contrast to the Strong Anthropic Principle, which according to Barrow and Tipler, holds that 'the Universe must have these properties which allow life to develop within it at some stage in its history' (1986, p. 21). The Weak Principle appears to rest on the truism that we cannot exist in a universe which precludes our existence. Yet a physical theory that could not allow for the creation of stars would be just as deficient, since physics has to explain things as they are. The Weak Principle does stress, however, that explanations cannot be produced piecemeal, because physical reality, as a whole, is linked with the history of the universe from its first moment. Indeed, because our own immediate world is as it is, we can come to some conclusion about the level of fine-tuning needed to produce it. This, though, does not explain the fine-tuning. We might, then, say, as the Strong Principle does, that it had to be like that, and that there is something necessary about the physical constitution of the Universe. Everything depends on the notion of necessity being employed here. Physicists trying to provide a complete explanation of everything may well be attracted by such a view.

Much follows from the simple fact of the existence of human observers. What is not so clear is why the universe *has* to be as it is or why we *have* to be able to observe it. We cannot coherently talk of the universe being fundamentally different from the way it is, as long as we envisage ourselves inhabiting it. There may however be an infinite set of other possibilities out of which the particular conditions arose which made our life possible. There could be different universes, or different physical constants and laws of nature. Our visible universe is one of an infinite range of possibilities, including all possible initial conditions, physical laws and constants. It is supposed that these do not merely remain as possibilities, but that they are all realized. A hypothesis of many worlds makes it appear unsurprising that we and our universe do exist. If our existence is possible there is bound to be at least one actual universe, it seems, which can produce us. The harmony that appears so remarkable at first sight between conscious observers and the universe may in fact be untypical. Alternative worlds need not include observers who would be able to wonder at their apparently privileged position.

The multiplication of universes which by definition cannot be causally linked, goes beyond the bounds of ordinary physical theory. We could never obtain empirical confirmation of such distinct worlds, since each has developed independently of the others. The theory is in every sense metaphysical, and it is tempting to reach for Ockhams's razor, which is always at hand to discourage us from multiplying superfluous entities. If the engendering of an infinite number of unknowable universes is not in this category then nothing is. Yet physicists, following Guth, have argued that so-called inflation occurred after the initial singularity of the Big Bang, under which the Universe could have been divided into countless, separate domains. One analogy would be many inflating bubbles. This means that the universe visible to us may not be typical of the whole, and that the physical laws we assume are universal and may only be local in character. It is as well that proponents of the idea of inflation have been able to reassure their colleagues that it would not be possible to create a universe in the laboratory (Farki and Guth, 1987, p. 150). The reason given is that the outward velocity for unlimited growth is so large that it could only have originated from an initial singularity. Yet the very idea of scientists 'creating a universe' is curious. If the term refers to everything that physically exists, there could only be one universe, even if it had sub-regions, when different physical laws hold. The idea, though, of regions disjoined from ours in every way even in the one so-called 'Universe' is still an unsettling one.

One reason why some have speculated about the existence of other universes is, in the words of one writer, that 'the presence of vastly many universes, very different in their characters, might be our best explanation for why at least one universe has a life-permitting character' (Leslie, 1989, p. 69). The question, then, is whether the proliferation of such entities helps us to understand why we exist in our own particular universe. Even if we accept, as some logicians do, that all possible worlds exist, that does not mean that all possible worlds are actualized. They still only exist as possibilities. Otherwise they would no longer be merely possible. However hard we try, we cannot escape the question why our particular universe, with us in it, exists as it does. The idea that all possibilities actually exist seems as extravagant as the 'many worlds' interpretation of quantum mechanics (introduced by Hugh Everett; see Trigg, 1989, pp. xvii ff. and chapter 6) which generates alternative realities on the basis of the mathematical formalism used in quantum mechanics. All such approaches tend to produce cosmological and physical theories on the basis of mathematics without any metaphysical constraints. A fundamental metaphysical insight must be that we live in one world. There is one reality. Whether physical laws and constants have a universal validity or a more local application is a further question. The idea of the inflationary Universe in effect gets around the point of there being only one Universe by postulating an ensemble of 'universes', which are part of the same Universe, because they stem from the same origin.

Once we realize that there are not just many other possibilities, but also other regions of the actual Universe, where different conditions obtain, our unique form of existence may not seem so remarkable. This is not simply the profligate idea that all logical possibilities exist as actualities. It is a serious physical theory about the differences that may obtain within the one Universe, which may perhaps serve to undermine our wonder at our uniquenesss. The suggestion is that if our visible universe is not typical of the whole, its special features may not appear so remarkable. We could not live in a region of the Universe which is not conducive to life. Because, though other possibilities may have been actualized in other regions, it may not be surprising that chance has produced us. So, at any rate, runs the argument. The question is whether, given the realization of varying possibilities, that it is any the less remarkable that a universe (or region of the Universe) containing us actually exists. Given that our universe does exist, it must be a possible one, and if all possibilities are realized, it is then inevitable that our universe becomes actual. Our universe then has to exist, and the anthropic argument shows the link between us and its order. Once, though, it is acknowledged that logic cannot conjure existence out of nothing, it will seem a fundamental error to suppose that all possible universes will actually exist. The multiplication of universes in itself does nothing to explain our existence. As long as it is recognized that not all possible universes will come into existence, even if many do, we have to return to the question why *this* one with us in it does exist.

The mere provision of other universes should not lessen our surprise at the existence of one conducive to life. Universes have been envisaged in parallel, and also as part of a cyclical process, with others preceding and succeeding ours. Physicists argue about how far physical theory allows for, or perhaps requires, such possibilities. Since all such universes, though, will be totally distinct from our own we could never gain any evidence of their existence. To call such reasoning speculative is an understatement. Without illegitimately jumping from possibility to actuality, and assuming that whatever is possible must also actually exist somewhere, we could never decide whether they really do exist. Yet *if* they do, what does that prove? The anthropic principle shows that they could not contain life as we understand it if they had a different physical constitution. The argument seems to be, though, that their mere existence somehow explains the existence of our type of universe.

One reason for the postulation of different universes is that those who wish to explain the obvious existence of order in terms of chance cannot do so by depending on the characteristics of our observed universe alone. Some have thought that even in the physical universe we inhabit, a sufficient number of totally random events could result in human life spontaneously arising. The hackneyed example was always that given enough time monkeys could type the works of Shakespeare. The argument from fine-tuning shows that this kind of appeal to chance is not enough. Our existence is indissolubly linked to the conditions prevailing just after the Big Bang. Many chance events could have stopped human life appearing or have wiped us out as the dinosaurs were wiped out. Our evolution however, was closely bound up with the fundamental nature of the universe. Devotees of chance, rather than purpose, are pushed back to posit chance as the explanation lying behind the appearance of this type of universe.

Why, though, should the fact that there are other universes, which may not support life, make it more intelligible that we happen to exist? There must always be the question why this or that universe came into being, unless all possibilities are realized. The answer could be that it was a chance occurrence, but it is hard to see why a string of such occurrences help to reinforce the view that it has all happened by chance. A succession of chance universes would seem even more extraordinary than the existence of one. One cannot explain a mystery by producing more mysteries, even an infinite number. Whether we inhabit the only universe is irrelevant to the question whether it came about by chance. Why anything exists at all is a question that applies to multiple universes as much as to a single one.

The anthropic principle provides a vision of actual unity when it emphasizes the interlocking character of the various parts of our universe, and the uniformity of physical processes throughout its time and space. This meshes in well with the metaphysical picture of there being one world, one reality. The idea of a reality that contained many disparate universes is pushing this metaphysical principle into vacuity. If one world contains universes that cannot interact, in what sense is it *one*? If they do not even have a common origin, but are part of an infinite set of possibilities, the unity beloved of much metaphysics has been lost. One can say that metaphysics is concerned with whatever exists. If what exists is internally disconnected this may be something which metaphysics has to contend with. It is important that metaphysics provides a proper context in which physics can operate. Once physics is conducted in a total metaphysical vacuum then no doubt the wildest of speculations can be conducted under its banner. The more we imagine alternatives where the laws of our physics do not apply, the more we are taking leave of science. There are limits to the differences scientists can envisage. If a world diverged absolutely from ours, we could not even properly call it a *physical* world. In fact, we could not in principle ever obtain scientific knowledge of another universe. This makes it even clearer that postulating its existence must come from a motive which is itself non-scientific. One powerful one is the feeling that the very existence of a universe has to be explained. It might be tempting to think that multiplying universes somehow disposes of that question, but, as we have seen, that is not the case. Being profligate with universes is a very dubious metaphysical activity. Simplicity must be a virtue, particularly as one can never make testable predictions about other universes.

Our Place in the Universe

From the very beginnings of philosophy in the world of ancient Greece, metaphysics has shown a preference for the one over the many, for unity over diversity. This preference has perhaps been even more pronounced in the sciences. The desire for explanation inevitably propels us back from the multiplicity of different causes to consider the possibility of an underlying simple structure. Physicists and mathematicians themselves always prefer the simple to the complex, the elegant to the unwieldly. The truth can sometimes appear complex, but scientists try to weld their different theories together to make sense of the whole. It is regarded, for example, as a major disadvantage that physics cannot provide a complete and consistent theory in order to unify Einstein's general theory of relativity and quantum mechanics. As Stephen Hawking puts it, 'the eventual goal of science is to provide a single theory that describes the whole universe' (1988, p. 10). Postulating different worlds, which are inaccessible from ours, must undermine the main purpose of science. At best it would be reduced to an examination of phenomena that could be called local, in comparison with the rest of reality. Indeed, if the notion of alternative universes is pressed hard enough, the idea of there being grounds from within physical theory even for envisaging them becomes preposterous. We would be using the insights gained from the structure of our universe to make assertions, of an uncheckable kind, about universes to which *ex hypothesi* the facts about this universe are irrelevant. The exercise only makes sense in terms of envisaging different regions of one universe where despite local variations, the same physical laws apply at some fundamental level. That, though, is still to ground diversity in unity.

Science can only properly proceed with the urge to unify, and to ground divergent explanations in something fundamental which all science has to acknowledge. It needs the ideas that scientific laws are universal and that physical constants are invariant throughout the universe. The alternative is stark. An acceptance of the legitimacy of different, and possibly inconsistent explanations, means that there is little check on which scientific theories can be held. Unless it is believed that when two theories conflict one at least must be wrong, it would seem that anything can be believed with impunity. Much has been made in recent years in the philosophy of science of the notion of incommensurability (see Trigg, 1973, pp. 101 ff.). This holds that different theories will posit their own entities, with no neutral way of comparing them. It is easy on such a view to lose sight of the fact that there is one reality which all such theories are trying to describe, with greater or lesser degrees of success. The concept of incommensurability makes a radical challenge to the possibility of rationality in science. Once the idea of one world common to all scientific theories is given up, we are bound to realize that there are no real reasons for holding one theory rather than another. Many look to social science to provide the explanation. It seems that science without a vision of unity is in danger of collapse. Yet the quest for unification cannot rest on purely methodological grounds, but has to rest on considerations concerning the actual nature of the world. If it depends on facts about us as human beings, or on the structure of our minds, it is changing the subject. It must, rather, be grounded in metaphysical considerations concerning the nature of reality itself.

That is why the anthropic principle, to be effective, must be a principle about the character of the universe, rather than merely one about us. It may start with the unexceptional fact that we are here, but it must not make the universe anthropocentric, let alone dependent on human minds in any idealist fashion. Hawking paraphrases the anthropic principle as saying: 'We see the universe the way it is because we exist' (1988, p. 124). This, though, can appear to put a wrong emphasis on the matter. If we did not exist, we would indeed not see the universe in any way. The important point is to stress the connection between the fact of our existence and the character of the universe. How we see it is a different issue. The fact that creatures like us are able to exist shows a lot about the character of the universe from its early stages. The function of our observations and reflection is to reveal this character to us, and to see the underlying links between such disparate events as the formation of carbon in the universe, and our own personal existence. Hawking's formulation moves our attention from what the world is like to the way we see it. As always, this may seem innocuous, since we could not refer to the world in a manner other than the way we 'see' it. The seeds of idealism are here. The anthropic principle can easily be coupled with interpretations of quantum mechanics which require the presence of an observer. The Copenhagen interpretation linked with the name of Niels Bohr particularly lends itself to some form of anti-realist conception of the sub-atomic world. The result can be a theory which uses the anthropic principle to link physical reality with the fact of our existence, and simultaneously makes observation by humans the criterion of reality. Wheeler has argued (in his introduction to Barrow and Tipler, 1986) that 'quantum mechanics has led us to take seriously . . . the view that the observer is as essential to the creation of the universe as the universe is to the creation of the observer'.

This way of envisaging our position in the universe looks surprisingly circular. According to the anthropic principle our presence on this planet is indissolubly linked with basic facts about the constitution of the universe. Yet the universe in turn is made to depend on our presence and our ability to observe physical reality. We are here because the universe is as it is, and the nature of the universe depends on our ability to observe it. Even if we were to accept a quasi-idealist interpretation of quantum mechanics, and to extend this to cover the whole of reality, this must undermine the anthropic principle, rather than reinforce it. We could not give a proper causal explanation of our existence by linking it to the fundamental nature of the universe, since the whole idea of such a nature is itself made to depend on our ability as observers. A genuine explanation should not lead us around in a circle, but must somehow ground us in a reality that is logically independent of us. As we saw, with Rescher's conceptual idealism, reference to causality does not mix well with many idealist conceptions. The impact of an invocation of a causal chain can be dissipated if it is made logically dependent on the powers of our minds. Instead, for instance, of explaining our existence, it becomes a reflection of it.

The anthropic principle should not be in any sense anthropocentric. Its message is that our existence is tied to the conditions present at the beginning of the universe. That does not mean that the universe somehow depends on us, nor that the whole purpose of the universe was to produce us. Modern science has always taken great pride in expunging reference to purpose and final causation, in Aristotle's sense, from its explanations. Physical theory merely draws attention to the long line of cosmic coincidences to which our existence is tied. It is when the Strong Principle is brought in that purpose may seem relevant. If it appears that the universe *must* have the properties which produce life, any property necessary for life is necessary for the universe. No universe could fail to contain such properties. It looks as if the universe *has* to exist, and one may wonder why. Yet even here purpose may seem irrelevant. If things have to be as they are, what need is there for further explanation? It may be that a contingent universe would be in ever greater need of explanation. If things could have been different, we might wish to show why they have turned out as they have.

Barrow, talking of a universe which *has* to possess the form it does, says: 'All the presently unexplained values of the fundamental constants of Nature would, in such a unique scheme, be forced to possess no possible arbitrariness' (1988, p. 360). This would be opposed to the view according to which the present universe results from an initial set of random events. In that case, there would be possible universes, where life could not develop. Certainly if this universe is the only possible one, and the provision of life is built into its fundamental constitution, we might seem to have arrived at the Strong Anthropic Principle. Alternatives are ruled out. Yet does this prove that 'there is only one type of universe that is logically possible' (p. 360)? That is the way Barrow describes one of the options available to cosmologists. The matter is closely connected with the search for a Theory of Everything, but that should be a search for what is physically possible. *If* there is a universe, it may be shown that it has to possess the character it does. We can easily take it for granted that we are here considering the nature of the universe, and so can easily slide into thinking that there *has* to be a universe. We would not be worrying about it otherwise. Logical possibility is not the same as physical possibility, and it is important not to blur the line between the two. For one thing, not all logical possibilities are physically possible. Sliding from one to the other can also be coupled with the confusion between real possibilities and what is actually the case. It can then be supposed that if there is only one logical possibility, it is somehow logically necessary, and must therefore possess physical existence. The universe has been conjured into being.

A contingent universe need not have existed, and even if only one type of universe is conceivable, that does not prove that it has to exist. It can still be a contingent fact that there is a universe, even if, given that it exists, it must possess the characteristics it does. However impressive a mathematical demonstration, there must always remain the question of whether it is physically instantiated. Mathematics could claim truth even if there were no physical world. Given that there are mathematicians producing theories, the question whether there is such a world has already been answered. It might be said that that is the point of the Strong Anthropic Principle. The fact that we are here to enunciate it proves, it is claimed, that the universe has to exist, with us in it. It might, however, be alleged that what it shows is that the universe has to exist *if* we do, and that is not a major discovery. Our existence is interwoven with that of the universe which produced us. Yet our own existence is certainly contingent. None of us need have existed, and neither need the universe. There may appear something odd about our finding it conceivable that we had never existed. We could not reflect on the nature of the universe without existing. The Strong Principle may trade on some such thought, but that does nothing to rule out the fact that we might never have existed. Much can follow from the fact that we do exist, but nothing can make that existence anything but a contingent matter.

Anthropic Arguments

We have seen that anthropic arguments are liable to be linked to the question of a complete explanation. If our existence is somehow part of the reason why the universe has the character it has, it must figure in a Theory of Everything. Yet this is not thought satisfactory by a physicist such as Steven Weinberg, whose aim is to provide 'one simple connected set of laws' which do not need in their turn to be explained in terms of anything further (1993, p. 41). Weinberg is pursuing the possibility of a 'final theory' and acknowledges the possible relevance of anthropic arguments, particularly if the existence of many different universes is envisaged. Our existence would at least show something about the character of the only universe in which we could expect to find ourselves. He hopes, though, that in the end, our own existence will not play an important part 'in explaining why the universe is the way it is'.

He continues: 'As a theoretical physicist, I would like us to be able to make precise predictions, not vague statements that certain constants have to be in a range that is more or less favourable to life' (1993, p. 182). He is not enthusiastic about appeals to anthropic considerations and, instead, wants a final theory 'that will turn out to have enough predictive power to be able to prescribe values for all the constants of nature, including the cosmological constant' (p. 183). Only, as he says, 'if all else fails', does he want to look to anthropic explanations. From this perspective, the latter seem very much second best to a reductionist approach which looks to a single source of explanation for the existence of everything. This would have to be couched in mathematical terms, which are far removed from reference to humans.

All anthropic arguments have the characteristic of reasoning backwards from our present situation. This is perhaps why they seem to prove more than they possibly can. Because they take the present state of affairs for granted, they can merely argue for what had to be the case to have produced it. Something is lacking in this kind of proof. It is not like a typical argument about the perils of self-reference and reflexivity. That would warn us against denying the truth of what has to hold to enable us to make the assertion in the first place. Anthropic arguments warn us against a picture of the universe which leaves no room for explaining how human beings were produced. Yet this, unlike self-reference, is not a question of logic. It is not about statements contradicting each other and thus about logical impossibility. It is about physical impossibility and possibility. If we exist as physical beings, a physical theory making that inexplicable is not very useful. Anthropic arguments have accepted existence as a premiss, and so have already taken something for granted. They need never confront the question why anything exists. In fact, anthropic arguments give the illusion of explaining more than they do. Concentration on the fine-tuning of the universe needed to produce us still does not explain our existence. The length of a shadow cast by a building at a particular time is inextricably linked to its height. No-one, however would imagine that the shadow's length explains the building's height. In the same way anthropic arguments, with their retrospective reasoning, establish connections, but still fail to give any explanation of why the universe exists or why we do. At least these who talk of multiple universes do see there is a *prima facie* demand for explanation, and try to block it by appealing to chance. A different reaction to the same problem is shown by theists who see evidence of deliberate design in the remarkable physical coincidences that lie at the basis of the universe.

It is sometimes alleged that anthropic arguments are too metaphysical, and that, as Weinberg suggests, cosmic coincidences might one day be given an ultimate physical explanation by scientists as yet unknown. The Strong Principle may give the illusion of appealing to some form of metaphysical necessity, but we have seen how anthropic arguments draw attention to the unity of the universe without really explaining it. Many find objectionable the way in which human beings are made to appear to be the linchpin of the universe. Even if we lower our sights from the heavens, and confine ourselves to the contents of this planet, why should we pick ourselves out as the most important inhabitants? Could we even look at inanimate objects and produce, say, a refrigerator principle? Given the existence of refrigerators and the materials from which they are constructed, the universe may seem to have to be as it is. An immediate reply would be that they are artefacts which depend on human purposes and intentions, and so their existence has to point us back to the fact of our own. There could, though, be, particularly for dog lovers, a canine principle which ties the existence of dogs in to the very structure of the universe. Dogs do not depend on humans in the sense that refrigerators do, and at least such a principle could serve to remind us that it is the existence of life as such and not just human life which is being tied to the development of the universe.

When, however, we look more closely at anthropic principles, an implicit message is that it is not just our existence being linked to that of the universe, but our understanding and consciousness. A connection is made between our ability to make sense of the universe and our presence in it. In other words, humans are being given a privileged position because of their rationality. In that case, a canine principle would miss some of the point. Certainly cosmologists like Wheeler are above all interested in our consciousness of reality and not just our presence in it. In other words, the anthropic principle is part of a wider search for an ultimate explanation, and, moreover, one which is intelligible to us. Whether we can ever attain to such an explanation is one matter. Denying its possibility is much more serious since that implies that in the end reality is not intelligible or comprehensible. In that case, we must either give up a search for explanation, or be content with whatever is the cultural fashion of the day.

Anthropic arguments attempt to transcend the relativities of time and place by tying in the existence of the human mind and its ability to reason with the inherent rationality we seem to observe in the world. The arguments must fail if it is claimed that such inherent rationality is illusory, or, in conceptual idealist fashion, a projection of the human mind. In that case, they do not attain a grip on the real world. The anthropic principle can be combined with conceptions of evolution, describing in the first instance an evolving universe which eventually produced us. It shows how a certain level of order from the very beginning of the universe was a precondition of our existence. One answer to the question why the universe is orderly is precisely that it is because we are here. We could not live in a chaotic world, or in a universe which did have a basic regularity in its processes. Yet as we have seen this kind of explanation explains very little, since it takes the fact of our existence for granted. It certainly fails to show why we should understand the order which has produced us. Plenty of creatures survive more or less successfully as products of evolution without being able to understand the situation.

Why should we be able to understand something of the unity and order on which our own existence undeniably depends? It seems odd to suppose that evolution *had* to produce anyone with the intellectual power to reason about the very nature of things. A disordered, unstructured, perpetually random world would not have been the kind to produce living creatures. The fact that we are alive proves something about the nature of the universe, but it does nothing to explain the fact of our consciousness, or our ability to reason and find the world intelligible. It might seem easy to argue that we could not live in a disorderly world. Therefore we have to be living in an orderly and intelligible one. Therefore we can understand it. The jump from order to our understanding is too great, without some further explanation as to why the systematic regularity inherent in the universe is of a kind that can be understood by the human mind. Why do we, and not other animals, possess rationality? Our attunement to the world in which we evolved may seem unremarkable. Unless we were so attuned, we could not have survived. Yet beyond a certain point the apparent ability of human rationality to grasp hold of the ordered structure inherent in things should be remarkable.

Anthropic arguments may point to the mystery of the source of the higher levels of human reason. They should also undermine many idealist interpretations of the rational structure of things. At any rate, they could not work at all if it is maintained that the rational structure of the world is somehow a human projection. Their whole point is that the order necessary for the development of life is objectively real, and able to give some kind of causal explanation for the production of the conditions necessary for life. If rational order was merely imposed on a formless world by the workings of the human mind, it would have been impossible for any life to have developed. Order is not a reflection of our own rational capacities. It is a precondition of our existence. The world which we live in must be structured in a way which could be understood by a rational mind. Otherwise we would not be living in it. Just why creatures like us, with the requisite rationality, have evolved is a further question. One thing, however, is clear. Reason needs something to reason about. Just as we could never have lived in a universe totally subject to the vagaries of chance, so human rationality is only possible, because it is able to reflect, however partially, the order that is present in all things.

7
Science and Society

The Sociological Turn

For all its deficiencies, the anthropic principle seems to point to order and structure in the physical universe. Otherwise we would not be in a position to reflect on its nature. This roots human beings in a reality of determinate character. The argument makes large metaphysical claims because of the fact of our existence. Yet it only works on the assumption of an objective world independent of our understanding. This runs against a prevalent view in philosophy that metaphysics is, to say the least, unnecessary, and that we inhabit a plastic, malleable environment. The world, it is said, has no fixed character, and we can have no fixed method of recognizing it. It seems to follow that when science produces something like the anthropic principle, it is not discovering an important fact, but is much more in the business of construction. Yet the formation of science seems very puzzling if the world is so indeterminate that it has to be moulded by us rather than act as an external constraint on our reasoning. How can science proceed, and what is it about?

The anthropic principle comes from within science and cannot provide a basis for validating the whole scientific project. It is indissolubly linked with current physical theory. Yet this illustrates the way in which physical and metaphysical reasoning are interlocked. Metaphysics cannot ignore the claims of science,

and science cannot proceed without making metaphysical assumptions. The biggest issue is that of the alleged plasticity of
the world. Has the world a fixed character and indeed, if it has,
are we in a position to know anything of it? Everything may be
indeterminate, so that there is nothing to know, except, perhaps,
that things are indeterminate. In that case, science would perhaps
be creating pictures, like an artist painting. Indeed these pictures
would not then be *of* anything at all. There would be nothing
against which they can be measured. To take another image,
beloved of post-modernists, we could say that science constructs
narratives or stories. This is a view that has received encourage-
ment from the demise of empiricism, which has resulted in the
undermining of foundationalist theories in epistemology. They
attempted to provide a firm basis for our beliefs, and seemed to
provide a way of justifying the claims of science in a non-circular
manner. Now, as we have seen, scorn is poured on all attempts
to stand outside science in order to justify it. One writer says:
'This project now seems to me merely a modern, secular version
of the medieval project of providing philosophical proofs of the
existence of God' (Giere, 1988, p. xvii). The analogy is instructive, since whatever one's view of such 'proofs', religious faith
can certainly not survive a denial of God's existence. It needs
metaphysical presuppositions about the character of reality, even
if it cannot provide a cast iron guarantee of their validity. In the
same way, science needs the possibility of grounding, even if it
cannot give an absolute guarantee of any particular claim that it
is actually grounded in reality.

We have already seen how the pragmatist emphasis on where
we stand now can encourage a sociological explanation of beliefs in terms of particular practices. Criteria for knowledge can
easily be taken to be local, rather than universal, a matter of what
happens to be accepted in a particular society, rather than of
what ought to be accepted everywhere. Adherents of post-
modernism have encouraged this development. Increasingly, the
simple alternative to a foundationalism which based our beliefs
on basic facts about experience, has seemed to be relativism.
When it has appeared impossible to rest knowledge on an incorrigible base which everyone ought to accept, it has been easy to

accept at face value the many differences of belief that occur, even within the bounds of the physical sciences. Metaphysics, and any allusion to reality, is ruled out completely. Faced with disagreement about the character of the physical world, many are content to turn to social explanations, without feeling the need to adjudicate between different opinions. This attitude has spread from the exotic world of social anthropology to the more humdrum routines of laboratory science. In both, differences in belief are traced back to differences of social context. No attempt is made to measure them against some objective reality. The impossibility of a God's-eye view is accepted, as is the impossibility of extricating oneself even partially from the presuppositions of a particular society. Yet, as we shall emphasize, there is always an inherent instability in such a position. If everyone is so firmly rooted in a society, however large or small, how can sociologists extricate themselves sufficiently to recognize differences between societies, even if they do not wish to make normative judgements?

Many have simply changed the epistemological category of certainty into a sociological one. Certainty can be produced by the shared assumptions of a society, so that we are just conditioned to take some things for granted. What we are certain of is just what we happen at any given time not to question. Empiricism and pragmatism had already paved the way for such a view by their attacks on metaphysics. When they were in turn overturned, there seemed no defence left against the transmuting of epistemology into sociology, from a normative discipline showing us what kind of beliefs we ought to have, to a descriptive one showing what beliefs are actually held. This process was helped on its way by the work of the later Wittgenstein, T. S. Kuhn and the proponents of the so-called 'strong programme' in the sociology of knowledge, according to which knowledge becomes an exclusively social category (see Trigg, 1989, chapter 5). Science then becomes intimately related to the local norms of a particular kind of society. It then becomes natural to talk of *Western* science, and indeed the very concept of science becomes less fixed.

If 'science' is logically related to particular, changing conventions of what is and is not acceptable to believe, the character of science is bound to develop as the nature of science changes. It cannot any longer be understood to involve a fixed methodology, through different historical epochs or in different societies. Incidentally, this has the result that science can no longer claim any particular distinction because of its methodology. Instead of being an effective or the only possible approach to reality, it is seen as merely an intrinsic part of the society we live in. Its authority has to be understood in political and sociological terms. Scientific method is valued because of our social background. It gives us no special lever on truth, since that cannot be identified separately from the conventions governing what is acceptable in a particular society. The sociological turn of much contemporary philosophy ensures that we concentrate on the fact of a belief or set of beliefs being held. Once metaphysics is removed from the scene and epistemology undermined, it is impossible to deal with any question of their purported truth. All we are left with is the role these beliefs play in a society.

In this kind of situation, falsity becomes mere deviation from a social norm, and ontology has to be discarded. Reason can no longer be understood as a faculty which can somehow transcend social boundaries in order to ascertain truth. Indeed, one sociologist dismisses all logic and reasoning on the grounds that they form the '*post hoc*' rationalization for ordered practices and conventional ways of proceeding. He says:

> Forms of logic, rationality and reason are then formal statements which reflect our acceptance of institutionalized practices and procedures. They are the vocabulary through and within which we re-assert the primacy of consensual practice and institution. (Woolgar, 1988, p. 48)

Reason is thus left with little role, embedded in social practices and used to re-assert their prejudices. It is mere rationalization, affirming what is already accepted in a social context. It can indeed have no focus as all objects are themselves to be regarded as socially constructed. The very distinction between an object and its representation is abolished. Science is indeed charged with having a so-called 'ideology of representation' which wrongly suggests that 'objects' 'underlie or pre-exist the surface signs which give rise to them' (Woolgar, 1988, p. 99). This insistence that there can be no distance between representation and object, or, as others put it, between subject and object, means that our theories cannot be distinguished from the world. Our writings cannot be separated from what we are writing about. They are logically related. We are told that 'the organization of discourse *is* the object' (p. 81). Science, in fact, is in the business of construction. It is claimed: 'Scientists are not engaged in the passive description of pre-existing facts in the world, but are actively engaged in formulating or constructing the character of that world' (p. 87). Science is essentially social not just in the obvious sense that modern scientists typically work together in teams. What science is about is 'the sum total of the operations and arguments of other scientists'.

It follows from all this that so far from 'nature' and 'reality' being the preconditions of scientific activity, they are its by-products. What really matters is the agreement reached by scientists at any given time. Science does not even pretend this to be a body of knowledge or would-be knowledge, since it is simply the product of quasi-political activity amongst scientists. It is argued that 'negotiations as to, say, what counts as a proof in science are no more nor less disorderly than any argument between lawyers, politicians or social scientists' (Woolgar, 1988, p. 89). Yet it is one thing to demonstrate the continuities and analogies between scientific reasoning and other forms of argument and persuasion, and quite another to use this point to undermine the possibility of reasoning towards truth. The shadow of positivism still lies over us if we assume that the only alternative to having empirical foundations firmly laid is for scientific reason to be displaced by power struggles and other forms of social 'negotiation'.

Deductive reasoning from uninterpreted, raw data is not the only model for reasoning in science, and as analogies with legal disputes should suggest, rational argument is possible without it. An argument between lawyers in court may be a display of rhetoric and the arts of persuasion. Plato was well aware that a clever lawyer could make a jury believe what he wanted. That, however, is not the purpose of the court, which, in a criminal trial, is to determine what happened and whether the defendant is guilty of the charge being put. Witnesses are called to try and establish the facts. Arguments and cross-examination are pointless unless it is assumed that truth is at stake. In fact, if the rival claims of scientists and their reliance on empirical evidence is regarded as similar to a dispute between opposing barristers in court, this does not undermine the rationality of science. Sociological analyses can be given of legal processes, and the art of rhetoric can help us understand the subtle methods of persuasion that can be used. Nevertheless these are pointless, unless we realize that the basic purpose of the courts is to discover truth as best they can in the light of the evidence available. The latter may be interpreted in different ways and some may conflict. Witnesses can lie, but, more important, people tend to see things differently even when they are present at the same event. The post-empiricist emphasis on the impossibility of uninterpreted data will be fully borne out in any court of law. Anyone who has listened to a succession of witnesses to a road accident will recognize that each person, however truthful, will select and emphasize different aspects of the event. For one thing, they each saw it from a different position. Courts explicitly recognize this and one of their roles is to challenge and test alternative interpretations.

Whatever theories may be advanced, it must still be the case that either the defendant stole or did not, that either someone killed or did not. It might, for instance, be proved that someone was not even present at the time and place of an alleged crime. In that case, the person must be innocent. It may be that 'proofs' in science may be more like a proof in a law court than a proof in mathematics. Scientific theories about the real world may aspire to truth by aiming for the deductive validity of a logical proof, but this is an impossibility. The very first description picked as a premiss will already be far from neutral, and may itself be open to challenge. That does not mean, however, that it is not subject to the constraints of the real world. It may not be as easy as Popper envisaged to falsify a theory. An apparent falsification may be better met by modifying a theory rather than abandoning it. Nevertheless it can become painfully obvious that a theory is failing to deal adequately with the world. There are cases in science to match the alibi which proves a defendant was innocent, as when a definite prediction is falsified. Not all instances, are as neat as finding a black swan when it was maintained that all swans are white, but they can arise. If they did not, the world could never catch us out as it so patently does.

The Sociology of Scientific Knowledge

The mistake of many sociologists of science is to suppose that once science has been knocked off its pedestal and shown to be comparable to other forms of social agreement, involving argument, then ideas of truth have to be abandoned. Yet social science is as dependent as any other form of intellectual activity on the idea of truth, and on the separation of subject or representation, and object (see Trigg, 1991). The sociology of scientific knowledge takes as its focus the work and assumptions of scientists. At least at first sight, this would seem to involve a distancing of the social scientist from what is being studied. Theories about the social context of science will be put forward which purport to be about something independent of the social scientist. Sociological descriptions have been substituted for those of science. Instead of allowing that physical scientists attempt to deal with a real world, some sociologists give their own analysis of what is going on, in a way that makes the physical world inaccessible and redundant. They give sociological descriptions of the interplay between scientists, the social conventions that govern them at work, and the way in which a picture of the world is constructed. Yet these descriptions gain their power and significance from the assumption that the sociologist can be independent from what is being described. The basic format of subject and object, theory and reality, is being transposed from the relationship of scientist to world, to that between social scientist and the community of scientists.

Whether social science can itself survive the repudiation of the distinction between analysis and object is a crucial question. In any piece of sociology it seems that sociologists have to adopt the external 'objective' viewpoint which is said to be typical of the scientist. When we are told by some sociologist that scientists do not discover what is real, but produce it, this is offered as a claim which gains its force because it purports to apply to something independent. Scientific activity is assumed to be a fit object of sociological study, and hence implicitly to possess its own reality independent of the study. Yet this means that sociology has to presuppose the kind of detachment from an objective reality, which it is simultaneously saying is impossible for the physical sciences.

Why should sociologists ever wish to deny that physical scientists are making discoveries? If everything is a projection of society, none of us, whether scientists, sociologists or anything else, will be able to engage with the world. Yet this does not prevent many so-called 'social constructionists' from viewing society and not nature as the source of scientific theory. There is, of course, a sense in which they must be right. When sociologists concentrate on the social conditions necessary for the construction of theory, they see much that is important. The whole background of laboratories, complex instrumentation, social networks linking scientists, the ways research findings are disseminated, and the role of public finances, to mention just some elements, can easily be forgotten. The simple positivist model of individual scientists deducing theories from their experience misses almost everything that seems crucial for modern science. A reaction against it, and an emphasis on the role of shared theories, can produce the conclusion that belief in theoretical entities is the end product of a complex social process. Many groups of scientists have interacted in ways that are sociologically explicable. For example, authority, as a social category, has its place in science, and can be very effective in blocking the publication of outlandish new theories. One of the seminal papers on the theory of chaos, entitled 'On the nature of turbulence' was originally refused publication, and in the end was only published in a journal of which one of the authors was editor (Ruelle, 1991, p. 63). Admitting he had to accept the paper himself for publication, he commented: ' This is not a recommended procedure in general, but I felt that it was justified in this particular case'. That kind of constraint on the publication and dissemination of new theories is grist to the sociologists' mill. Their aim is to show how social terms alone are adequate to explain what scientists come to accept.

Yet many will protest that even if social forces facilitate or obstruct the production of new scientific knowledge, what makes it knowledge is its relationship with the world and not its role in society. This is precisely what many sociologists of scientific knowledge deny. For them, knowledge is a social category and not an epistemological one. They would repudiate any idea that rational and social factors could be distinguished as the determinants of the adoption of a scientific theory. It is not, they would say, a matter of appealing to logic or sound evidence, on the one hand, and social causes on the other. That would be a massive begging of the question. One writer puts it this way: 'The question they beg lies in the assumption that "logic" and "sound evidence" are not in themselves anyway "social". It is just this sort of assumption which recent work in the Sociology of Scientific Knowledge has challenged' (Pinch, 1990, p. 659).

There are various staging posts along this path. It is possible to deny that cognitive pursuits and their social organization are two independent entities (Fuller, 1988, p. 9). In that case, the norms of epistemology cannot be kept separate from the ways in which the pursuit of knowledge is conducted. A 'social epistemology' is not then a contradiction in terms. Many sociologists have, however, much greater ambitions than the blurring of classical boundaries. They want to give full-blooded sociological explanations of every notion that was formerly within the province of epistemology. This can be done quite easily if it is accepted that notions such as 'evidence' and 'reason' have no independent life of their own. Good evidence is then merely what is accepted as good. Evidence is what is thought relevant to a particular issue. Reasons are what are considered to be grounds for something. All these epistemological notions would appear to carry with them a tacit reference to being accepted. It can appear that nothing has validity in the pursuit of knowledge unless it is seen to have it. It is all too easy for a sociologist to seize on this and claim to provide a social explanation why a community accepts one thing as evidence but rejects another an irrelevant.

The problem with all this is that epistemology is not a descriptive discipline, showing us what is accepted. It is a normative one telling us what is acceptable. Its conversion into a branch of sociology must involve the loss of its most distinctive feature. Yet what then is the source of its ability to guide us as to what to believe? The sociologist would reply that this can only lie in the conventions prevalent within a particular society, and that these are not rationally grounded. Social norms must be understood in sociological terms. This process continues right into the heart of scientific practice in the laboratory. We are told that 'laboratory studies display scientific products as emerging from a form of *discursive interaction* directed at and sustained by the arguments of other scientists' (Knorr-Cetina, 1983, p. 118). Social constructivists explicitly denounce any idea of science as being *about* the world. Scientific 'products' are regarded as being above all the result of a process of fabrication. According to the constructivist interpretation, 'the study of scientific knowledge is primarily seen to involve an investigation of how scientific objects are produced in the laboratory rather than a study of how facts are preserved in scientific statements about nature' (p. 118). The conclusion is unmistakeable. Scientific reality progressively emerges out of 'indeterminacy and (self-referential) constructive operations' (p. 135). There is no need to assume that it matches any 'pre-existing order of the real'.

Another writer in the area of the sociology of science pursues a similar theme. He says:

> What we need is radical uncertainty about how things in nature are known. The radical uncertainty is relativism. Any alternative view can be crippling to a vigorous explanation of the social construction of the rational world. (Collins, 1983, p. 91)

Any thoroughgoing theory of social construction must link scientific reality to social conditions, and since these vary, it follows that what counts as physical reality will depend on one's society. Yet at this point it becomes crucial to decide whether we mean Western society, a narrower segment of it, the community of scientists, or possibly communities of particular scientists. Since laboratory studies are so much in vogue it would not be too far-fetched to trace back the social construction of particular parts of reality to one laboratory. The social interaction alleged to produce 'scientific objects' could be of an exceedingly local kind. There would, of course, be the problem as to how new discoveries could be shared with scientists from other backgrounds. The temptation must then be to widen constantly the idea of a society or community, in step with the ability of some to persuade others to accept their novel constructions. The idea of a community of scientists worldwide may be attractive, but it is far from clear that a commitment to the standards of Western society would be sufficient to outweigh all the other distinctive social influences which undoubtedly have their effect on different people in different places. The more, too, that the concept of science is abstracted from its social background so that it can be applicable everywhere, the more is it being admitted that society is not the only determinant. Other criteria can be applied.

We can avoid these problems by refusing to jettison the notion of an independent reality to which anyone anywhere can have access. One group of scientists can draw the attention of others to features of reality which are invariant between societies, but which may not have been previously noticed. This is connected with the importance, universally recognized in science, of being able to repeat experiments. The constructivists would see this as a purely social exercise, a sharing in the process of construction and acceptance. They could not however, explain why it is in principle possible to repeat the same experiment in different social settings. We have a ready-made explanation for this if physical reality is a constant between societies. The constructivist would have to demand that the exact social setting be replicated. If, however, society means anything more than the ability to manipulate certain types of apparatus, this would seem unlikely. All interpretation of experiments will be understood as itself socially grounded, and so sense can only be made of 'findings', according to constructivism, by means of the social underpinning it sets out to explain. Even what is 'seen' in an experiment, and whether it is later treated as a fact or mere 'noise' are matters that are themselves 'embedded in a local history of technical activities' (Lynch et al., 1983, p. 220). Certainly decisions have to be made about the possible significance of what experiments show. Sometimes what is regarded as mere distraction and dismissed as of no account is later seen to be of enormous importance. This happened in the development of theories about chaotic phenomena. The mere fact of 'chaos' could seem to be an obstacle in the search for underlying regularities, and was often ignored. The question remains, however, whether what is noticed and what is discounted depends on arbitrary social arrangements to the extent that they are actually *creating* the phenomena, as opposed to, say, influencing people's reactions to them.

Social Convention and Reality

Sociologists often invoke Wittgenstein's conception of a form of life to illustrate the way in which scientific 'discoveries' are welded into a whole body of social interaction and practice, and cannot be understood apart from them. For example, theoretical terms are called 'constructs' and then said to be 'embedded in instrumentation networks, action and praxis, in short within a society or form of life' (Pinch, 1990, p. 661). The notion of a form of life, however, is no more helpful or less vague than that of a society or community. Indeed the concept seems to possess no function beyond pointing to pockets of agreement in judgements. Yet even here there is much vagueness. The practice of science, with its agreed procedures and methodology, might seem to make a promising candidate for a form of life. However, those ethnographers who enter a laboratory as if they were investigating the practices of an unknown tribe do not seem to be invoking the universal agreement of scientists as to correct procedure. The whole point of their work is to look at local practices, with their differing conventions and varying tacit agreements. The upshot must be a view of the fragmentation of science into different societies. As always, when the concept of a form of life is invoked, we are liable to find that basic disagreements will define the boundaries of a form of life. We will then be inexorably driven in a downward spiral to ever smaller groups, searching for a firmer spread of agreement. The trouble is that we could reach a stage, as in religion, where it appears that a form of life has only one member, on the grounds that there were important disagreements even between individuals. It would then be obvious that the concept is useless as a tool for explanation. The very notions of society and social conventions would have evaporated.

The stress on construction rather than discovery in science is remarkable in that it appears to go directly against common sense. A rejoinder to this could be that such common sense is actually the legacy of the philosophic views of previous generations. Sometimes sociologists make the grudging admission that scientists have to engage with a material world. One writer advocating a 'pragmatic realism' adds that such engagement 'is never unmediated and direct' (Pickering, 1990, p. 707). He says that 'it always has to be understood as conditioned by the culture within which it takes place and as reacting back upon that culture'. Our response to the material world is caught up in a set of hetergeneous elements that we are told 'sometimes come together in a loose and fragile unity one could call a "form of life"'. What began, therefore, as an external constraint on science is immediately down-graded in favour of a notion of form of life that is explicitly so vague as to be useless.

The more it is stressed that society, in one guise or other, is the mediator between ourselves and reality, the greater is the danger that reality drops out of account altogether. It will play no function in any explanation, particularly a sociological one. The more the social side is examined, the more it will appear to account for the content of our beliefs. How this works comes out clearly in an example canvassed by sociologists. The story runs as follows:

> Columbus 'discovered' America, but this discovery . . . did not *count* as knowledge (in fifteenth-century Europe) until it had been socially ratified. . . . And in a *certain* sense . . . America did not 'exist' until Columbus's knowledge claim was socially ratified, and thereby entered meaningful European discourse. (Oldroyd, 1990, p. 641)

Analogies are drawn between this case and that of science. The upshot and intention is to show how all knowledge is socially

constructed and to demonstrate that any idea of reality cannot
be entertained apart from what is accepted according to the
norms and conventions of a particular society. If all concepts are
rooted in society, it seems to follow that we cannot appeal to
what lies outside our society to explain our beliefs. It is however
fallacious to conclude that even if our concept of reality is the
product of society, so is reality itself. Concepts should never be
confused with the reality they attempt to refer to. The concept
of a dog is not a dog. Any suggestion that the social ratification
of claims to knowledge has any bearing on the reality that is said
to be known must be ludicrous.

A continent cannot be conjured into existence because of what
people in Spain or England chose to believe at the end of the
fifteenth century. Even more to the point those in a society could
believe something very fervently and be mistaken. The earth was
never flat even when everyone believed it was. Moreover em-
bracing relativism must be disastrous for the practice of science.
Scientists need to believe that they are investigating *something*. They
need to believe they are discovering truth and correcting error.
If they come to believe, along with the sociologist, that their
practices are simply complex social arrangements, with their own
dynamics, it is hard to see how they could continue with them.
One writer asserts bluntly that science is less a matter of truth
than of making worlds (Cross, 1990, p. 205). Yet only a few pages
before he had also expounded the persuasive view that belief in
mind-independent entities is important for scientists. Calling the
position 'motivational realism', he asserted its importance as
a regulative principle (p. 200). The possibility of having theories
about reality 'is the psychological anchor that makes a life in
science meaningful'. Yet there seems to be unbearable tension
between the idea of making worlds and of theorizing about reality. If scientists are actually doing the one while having to think
that they are actually doing the other, how can this be resolved?
The question is whether physical scientists can calmly continue
with their work, if they were to accept the sociologists' claims.

Questions of motivation cannot be completely separated from
people's own understanding. Once the metaphysical props of
science, and not least the idea of an objective reality, have been

removed, it is naive to think that things will go on as before. Most working scientists are realists at heart, and have an unreflective belief in the reality of the physical processes they are uncovering. If they were persuaded that this is an illusion, they might even decide to give up. An analogy is religion. It is hard to think that religious practices can long endure after people have given up any idea of the objective reality of God. Indeed anti-realist interpretations of religious belief are very similar to anti-realist revisions of our understanding of science. Changes in our understanding of the meaning of our practices, in whatever sphere, are unlikely to leave the practices unscathed.

Science as Rhetoric

Sociologists are quite likely to refer to 'the rhetoric of realism', and it does suggest that realism may have a very specific function (e.g. Woolgar, 1988, p. 106). It may be that scientists' apparent belief in an objective world is all part of the scientific apparatus. It should be understood not as a genuine metaphysical belief but as a rhetorical device, designed to persuade others of the claims of science and to uphold its authority. An unmasking of science as rhetoric would sit very easily with sociological efforts to describe the social basis for such an outlook. Individual scientists are not usually deliberately setting out to deceive, but it may be supposed that there has been a collective attack of false consciousness. The sociologist would hope to give the reasons for this. The project also combines well with a post-modernist emphasis on texts and their interpretation. Our attention is thus moved from the supposed reality described by scientists to the kind of language, which they use to persuade others of their discoveries. Indeed, as is pointed out, 'from the rhetorical point of view, scientific discovery is properly described as invention' (Cross, 1990, p. 7). The reason is that 'discovery' is in fact a hidden metaphor, which trades on the apparent reality of what is discovered. Cross writes: 'The term *invention* . . . captures the historically contingent and radically uncertain character of all scientific claims.' Scientists, it seems, 'invent' facts. Yet it is hard not to retort that this itself carries considerable rhetorical force. Anyone wanting to undermine the authority of science will find it very useful to persuade others that scientists invent what they claim to be true. The rhetorical analysis of science must itself be highly rhetorical.

Rhetoric is not presented as a powerful way of leading people towards, or perhaps away from, the truth. It is no longer seen as an instrument which can be put to good or bad ends. It is all that there is. Like some sociologists, Cross acknowledges that the 'brute facts of nature' exist, but he immediately discards them since he argues that 'only through persuasion are importance and meaning established' (p. 4). He says 'As rhetoricians we study the world as meant by science.' This, though, is very far from 'nature'. The range of facts worth investigating, how they are investigated, and the meaning of the results are all determined by rhetorical processes. The reference to nature betrays an understanding that, after all, science does have to be about something, but the concentration on the processes of persuasion cuts the link with nature. Even the idea of science as the paradigm of rationality can be seen itself as a clever ploy in the art of persuasion. Cross comments: 'The objectivity of scientific prose is a carefully crafted rhetorical invention, a non-rational appeal to the authority of reason' (p. 15).

When appeals to reason are dismissed as rhetorical, having the single aim of persuasion, the underpinnings not only of philosophy, but of any intellectual discipline are shaken. It may at first sight seem attractive to question the apparent arrogance of science in claiming a privileged position. Why should it not be just one intellectual enterprise amongst others? Why should it claim any superiority over philosophy, literary criticism or even rhetoric? The problem is that all attacks on the possibility of reason prove too much. If appeals to reason are rhetorical in science, they are likely to be rhetorical elswhere. Once *all* language is seen as an exercise in the art of persuasion, even if in different ways, no intellectual discipline can remain safe. The very means used to humble science can become a way of unmasking any discipline as a mere ploy to persuade others. What, though, is the point of such persuasion? Scientists may make illegitimate claims to objectivity, and pretend that they are talking about something they have not themselves constructed. It seems important, though, that they do not realize what they are doing.

Rhetorical analysis does not merely restrict the authority of science, or place it in a more modest position. It can destroy science. Once scientists realize that they are not even purporting to talk about anything independent of their discipline and its social setting, they are unlikely to want to persuade others of their so-called 'discoveries'. They will not even be able to persuade themselves. Any discipline that conceives of itself as entirely an exercise in rhetoric is hardly likely to go on as if it believed it were teaching truth. This applies as much to a subject like history as anything else. If it sees itself as merely concocting persuasive forms of narrative rather than describing and explaining actual historical events, its whole future as a separate discipline is put into question. History may indeed be novel writing but that is not how it is presently understood. Indeed there would seem little point in being skilled in the arts of persuasion as a historian, or a scientist, if there is no right or wrong, truth or falsity left. Science should not have monopoly claims to truth, or ownership rights to reality. Both concepts are, in various ways, fundamental to the exercise of human reason.

An historian who does not recognize a distinction between what historians assert and what actually happened cannot understand the basis of history. However much it is recognized that interpretation and facts cannot be separated, history collapses without some distinction between a correct and an incorrect interpretation. Even literary theory runs the risk of self-destruction once it allows that *anything* can properly count as a text, and *anything* can be accepted as an adequate interpretation. The allegation that language is being used in a merely rhetorical way will, in the end, stop us using it. The search for truth, at whatever level, has been transmuted into the use of power, but not everyone may be concerned with power for its own sake. Indeed the question cannot be shirked whether the claims of rhetoric themselves involve rational claims to truth, or are merely an ugly exercise in the power of persuasion. If they are unmasked as the latter, why should we take any notice? Rhetoric is certainly commended to us on apparently rational grounds. If its appeal is in fact a rhetorical one, designed to persuade, we need not, and indeed must not take any notice. Cross wishes to show how knowledge is possible without metaphysics (1990, p. 194). Knowledge is 'a matter of persuasion and consensus'. In fact, we are told that rhetorical interaction is constitutive of knowledge. However, as we have seen, one cannot destroy the metaphysical presuppositions of an activity, without changing the nature and purpose of that activity. An atheistic Christianity will be radically different from the traditional version. Science, when seen as a social activity, will be noticeably at variance with a science that is thought to be the pursuit of truth. Sociologists tend to become preoccupied with the fact of belief. They are interested in the interaction of people in a social setting. It is easy for them to forget that in any belief, whether in science or elsewhere, something is held to be true. If we become convinced that truth is an illusion, our beliefs are bound to be put in jeopardy.

Social Construction

One way in which social science typically varies from the physical sciences is that the fact that its theories are held and published can have an appreciable effect on aspects of society. The publication of physical theories will not change the behaviour of subatomic particles, but the publication of political opinion polls may certainly influence how people vote. Theories about the rhetorical function of science or its social roots could easily shake the faith of scientists in what they are doing. If such theories are being put forward as truths of general applicability, they must also be applied to themselves. Those who stress the social construction of reality must in all consistency accept that they too are constructing a reality, albeit one in which reality is constructed. We are back with the perennial question of reflexivity, which in fact is bound to afflict anyone who removes a metaphysical grounding for human reason. Without a focus, all claims would seem in the end to be claims about nothing. Assertions are no longer assertions about what is true. Yet the question is whether those who put forward theories of the social construction of theories can then go on doing so. Are they themselves able to entertain and advocate a belief which they know is a reflection of the society of which they are a member? They know that there can be no grounds for thinking it true, because that kind of truth is illusory. Why then do they advocate it? Many certainly glory in the problem of reflexivity and self-reference. Once though they recognize the nature of the social forces influencing them, it would seem difficult for them to continue to argue in good faith. The paradox is, of course, that even the recognition of their predicament involves a claim of truth. It appears that they see the reality of certain social arrangements. If these themselves are socially constructed because of other factors, we are involved in a vicious infinite regress.

Social constructivists must be driven by the implicit belief that social constructivism applied to science is *right*. It explains the way scientists interact. They have, it appears, discovered something about the workings of a significant segment of society. Yet their own arguments, if that is what they are, would suggest that this cannot be so. They have outlawed the word 'discovery', and are, like scientists, engaged in a social enterprise which as actors in it they may not have fully understood. There would seem to be as few grounds for their engaging in their activity as for anyone else taking notice of them. What they choose to construct seems in the end to be totally arbitrary, except in so far as they are constrained by the norms of their social background. Yet even that is to suggest that such norms can possess an existence independent of the investigation of sociologists. Otherwise they could have no causal power. Unfortunately many sociologists do wish to give causal accounts of the production of beliefs within a society. Social theorists who have rejected the idea of a rational grounding in reality for belief might well be very tempted instead to explain the origin of beliefs in terms of social practices. Yet the latter themselves possess a reality if they are to be causally efficacious. We seem not so much to have dispensed with ideas of reality or 'the world', as changed our minds as to what constitutes them. Cross argues that 'to leap from the social and linguistic practices of the laboratory to a universe of causally arrayed

physical objects is simply to mislay Ockham's razor' (1990, p.83).
Laboratory practices thus produce a belief in physical objects
and causal processes, and we are urged in the name of simplicity
to dispense with the latter.

It is very easy for anyone making a connection between beliefs
and social arrangements, to imply the existence of a causal link.
Some may rest content with this but the more radical are liable
to take the next step and say that because the one produces
the other, we have no rational grounds for believing in, say,
the objects posited by science. Yet the acceptance of the causal
connection depends on the idea of objective reality, of causation
being an independent process in the world. If all this is to be
understood as socially constructed, there is no way in which the
initial connection between social background and belief, between
laboratory practice and the positing of entities, can be properly
understood. Sociological analysis gains its authority by drawing
our attention to links between apparently disparate processes.
This works while a grasp on reality is maintained, but if the thesis
becomes a global one about the place of scientific knowledge in
society, and is then applied to itself, we seem propelled into a
downward spiral that can soon produce dizziness. If science is
socially constructed, so is social theory, and our theory of social
theory and so on. Only if there *are* causal connections, as opposed to mere beliefs projecting them, can any social constructivist thesis gain a purchase on the real world. There is a deep
incoherence about any attempt simultaneously to reject reference to the world as being a mere construction, and yet to
give an explanation of that construction in terms of causal processes which to do their job have to be independent of our
understanding.

Once realism has been repudiated, and the possibility of metaphysics undermined, it may seem that no room is left for a
distinction between the fact of a belief being held and its truth.
A relativist system can make a difference between the conventions
of the system and an individual's assertions within it. It is always
possible to break the rules of a game, even if it is only a game.
Those who wish to deny any distinction between subject and
object might, for this reason, be dissatisfied even with relativism,

and feel it has not been pushed far enough. Even if reality has been replaced by what is collectively accepted, a distinction between a particular assertion and its 'truth' can be maintained, in so far as it follows or deviates from the prevailing norms. Yet this situation is inherently unstable. It seems unlikely that when people claim truth, they are merely claiming solidarity with the prevailing opinion. They may indeed want consciously to reject the norms of their society. Prophets and rebels cannot always be ignored, simply because they are in a minority.

The problem is that, as a doctrine, relativism is a second-order one about the status of the beliefs of a society. It would not be held by the ordinary members of a particular society immersed in its institutions. Relativism only becomes possible when someone reflects on the nature of societies and tries to acquire a certain level of detachment. Historically, relativism has arisen when people are no longer members of a closed community, but are made aware of different societies and their different beliefs. 'When in Rome, do as the Romans do' is a maxim that could only arise when it is realized that not everyone lives in the same way. Non-Romans have to be aware of Romans. Belief in one religion becomes problematic when it becomes apparent that different societies have different religions.

A fateful step is made when the idea of truth is linked with the fact of acceptance. Even if individual deviations within a society are recognized, the criteria for truth adopted are logically linked with what its members agree about. Yet the problem always comes when relativism as a theory about the status of beliefs is applied to particular beliefs. What is to happen when relativists stand back from their own most cherished beliefs and wonder about their status? They will have to conclude that what they unthinkingly accepted as objectively true only possesses truth because they and other members of their society accept it. There can then be no ultimate answer to the question why anyone should have held such beliefs in the first place or go on holding them. Relativism not only undermines the possibility of justification but removes any point in believing anything, other than because one wishes to be a well-adjusted member of one's community. However, the idea of, say, believing in God not because one believes in God but because one wishes to conform, is rather curious. Social conformity and genuine belief have usually been distinguished.

If relativism *were* a correct account of the status of the varying beliefs of different societies no harm would be done until it was actually accepted. The inhabitants of a society cut off from the rest of the world could in all consistency take for granted that all the assumptions of their society were absolutely true. They would never be fundamentally challenged by the existence of alternatives. Relativism itself is the result of self-conscious reflection. It is the product of a rational process of thought which tries to grapple with the fact of disparate belief-systems.

One can only become a relativist by in a sense detaching oneself from beliefs. Yet this whole description of the situation only emphasizes the incoherence of any form of relativism. Reference to the possibility of a correct account or arguments in favour of a belief in relativism, presumably as true, show that all reasoned deliberation leads back to the distinction between truth and falsity, correctness and error. Without them there can be no point in believing anything, not even in believing the pointlessness of belief.

It may seem that science's demand for objectivity is just a feature of its own characteristic form of life. It could be given both a sociological and a rhetorical analysis, since each reinforces the other. The more that theories are said to arise from a particular social background, the more their content will appear irrelevant, and the use of the language expressing them will be scrutinized to reveal purposes other than communicating alleged truth. Yet this involves standing back from science and seeing it in conjunction with alternative forms of life. It involves comparison, treating different modes of human activity as existing in their own right apart from our understanding. It presupposes our ability to abstract ourselves from all of them and deliver pronouncements concerning their nature. The very notion of a form of life can only be derived from the kind of God's-eye view no relativist could accept. The very ability to acquire the concept suggests a detachment from ways of life and an examination of their independent existence. It suggests an ability to reason about a segment of objective reality, albeit social reality. The ability to entertain relativism as a possibility would seem to suggest its falsity. Doctrines about the relativity of truth presuppose a capability on our part of abstracting ourselves from all situations and declaring all equally defective. There is a sense in which relativism springs from the greatest arrogance of all, the assurance that nobody, anywhere, can be right, because somehow the relativist knows that there is no such thing as being right. Yet this is as much a piece of knowledge as any the relativist rejects.

8
Can Science Explain Everything?

The Scientific Millennium

A strong motive for resisting the attempts we have been looking at to give a sociological explanation of science is that they undermine science. They cannot allow it to give an account of the various processes at work in the physical world. Science may aspire to give explanations, but if these are in turn 'explained' by sociology, they can no longer be seen as being about physical events at all. They are merely reflections of prior social organization. Yet, sociologists are themselves seduced by the attraction of explanation. Otherwise, they would not try to show the social pressures at work in the production of science. We have seen how they, in turn, must be vulnerable to the same kind of treatment. Their theories will also be seen as the products of an antecedent social background. Ultimately, we must be forced back to something like a realist position if we are to believe that science (or sociology) can explain anything. Unless theorists can, in principle, gain contact with some reality, science can only be seen as a projection from society, or perhaps the human mind, and not as a genuine explanation at all.

Many scientists and others believe, however, not just that science can give some explanation, but that it can explain everything, presumably even the conditions of its own possibility. The idea of a final and complete explanation continually beckons. Yet, it is as well to realize that in that case there could be no possibility of science itself needing any justification. There would be no room for metaphysics, and no ground for searching for any knowledge beyond science. The desire for a complete explanation is crystallized, as we shall see in this chapter, by the search for a 'Theory of Everything'. Other developments in science, however, cast doubt on whether such a complete knowledge is in principle attainable.

Pragmatists always prefer to substitute the concrete for the speculative, and it is in this vein that Peirce looks to some final agreement between scientists. Yet, even here, it is hard to know how from within science we could view the agreement as final rather than temporary, fixed rather than about to shift. To see even this we would have to find a place beyond science to reason about it. The idea of completeness would not seem to be one which could arise within science. It is, in fact, not hard to see the desire for a complete explanation to be given in scientific terms as the product of a faith in science that has analogies with religious faith. Indeed, in so far as modern Western culture has a religion, it is science. The discoveries of science appear to govern our understanding of the world and our place in it. Above all, rationality appears to be defined by the empirical methods of the physical sciences, so that what cannot be proved scientifically must not be accepted. What lies behind the reach of the scientist in the laboratory cannot, it appears, count as knowledge. The very fact that 'scientia' (or 'knowledge') is restricted in the English language to a very specific understanding of a particular method of enquiry is a significant feature of our culture. This restriction of knowledge, and indeed of wisdom, to the experimental investigation of the physical world involves a remarkable narrowing of vision.

Theology's claims to be a science have been laughed out of court, while the social sciences have even been involved in political struggles to retain public recognition as sciences. This is not just a quirk of the English language, though the German 'Wissenschaft' retains a wider scope. It is a sign of a deeply rooted philosophical view in the modern world. Rationality is linked to knowledge, and knowledge to science. Science gives us the means for understanding reality and the rational person will act in accordance with that understanding. The conclusion, as one philosopher puts it, must be that 'the theory of what it is to be rational cannot be prised apart from the theory of science' (Margolis, 1986, p. 95). It is not just that it would be irrational to ignore the findings of science, but for most people the methods of science provide the standard for rationality. The investigation of the universe through observation and experiment is taken to be the examplar. Even though the role of theory has come to be explained more, the stress still seems to be on the importance of empirical discovery. The fear of metaphysics is endemic. Truth is equated with what is factual and the factual is assimilated to what is accessible to science.

C. S. Peirce wrote: 'Metaphysics is a subject more curious than useful, the knowledge of which, like that of a sunken reef, serves chiefly to enable us to keep clear of it' (1957, p. 56). Peirce was writing towards the end of the nineteenth century when it was easier to be optimistic about science and its links with human progress. Nevertheless his views, and others like them, have influenced the whole of twentieth-century thought. His emphasis was not so much on the nature of reality as on the ability of science to come to a settled opinion. The aim of inquiry, he thought, was the fixation of belief. What is real is what everyone will finally reach a settled agreement about. He says: 'The opinion which is fated to be ultimately agreed to by all who investigate is what we mean by the truth' (p. 53). Whatever may be meant by an ultimate agreement, Peirce wishes to subordinate the notion of reality to what scientists will one day think, or would one day think if certain conditions were fulfilled.

Pragmatism aims to link the notion of truth to the possibilities of actual inquiry. It is concerned with the world as *we* find it. Peirce would not accept a characterization of scientific reasoning which parted company with science as we understand it today and as he understood it. That is why he defines reality in terms of science as it will ultimately be, rather than science in terms of reality. He does not say that science will be complete when it has gained a full understanding of reality. His focus is on the actual activity of scientists and the knowledge it will produce if continued long enough. Knowledge is constituted by the agreement of scientists when it has become firm, as indeed, it could be alleged, is the case in many areas of science already.

The eschewal of metaphysics and the pragmatist emphasis on the actual procedure of scientists seems to make the vision of a completed science possible. Truth, reality and reason are then all linked to the aim of all scientists one day reaching a firm and settled opinion about everything. As a regulative ideal, there may be something to be said for this, although the reflection that in scientific matters most people have been wrong about most things most of the time during the course of human history does not inspire confidence. Certainly any theory of the nature of science which founded it on human judgement has to extrapolate to some ideal condition. Yet if it is still related to actual inquiry, the tendency of humans to fall into error is all too obvious. It was indeed a cardinal principle for Peirce that the scientist should always remember that his beliefs could be mistaken. He said of the scientist testing results: 'He stands ready to abandon one or all as soon as experience opposes them' (1931–58, vol. 1, p. 635). Once, however, we talk of hypothetical inquiry and a hypothetical final agreement as defining reality, the danger is that the whole idea underlying pragmatism is denied. Peirce tells us that 'we must look to the upshot of our concepts in order rightly to apprehend them'. If our concepts do not make any difference in the actual world they fail to gain a grip on anything. What effect, then on our actual world could the concept of a scientific millennium have? It is itself in danger of receding into the hinterland of metaphysics. An agreement that is an impossible ideal for people with our limitations is not one that should be entertained by any self-respecting pragmatist.

The notions of possibility and of conceivability are being strained once we begin to think in terms of all possible experience, and all conceivable evidence. What is experience if people like us could never have it? What could be conceived, if it is forever beyond the limits of human comprehension? Building up a picture of the world out of our actual capabilities and the actual possibilities inherent in human science is going sooner or later to come up against the simple fact of human finitude. We are limited in the scope and reach of our knowledge by time and place, and the kind of creatures we have evolved into being. Any urge to transcend such limitations must be metaphysical. It ignores the limits set on us and our descendants. In fact it is transformed into the simple view that if something is real, it must in principle be possible to know it. This, though, leaves un-answered the questions who would know it, when and where, and by what means. It is easy to pass imperceptibly from a ref-erence to present scientists to referring to future ones. Then, as that is clearly too limited, one could call in any rational inves-tigators, even if they belong to an unknowable species in the far side of the universe. That, however, still has its limitations and one can then call to aid omniscience, either in the guise of some abstract ideal observer, or perhaps of God. Yet even then there is a significant difference between defining reality in terms of God's knowledge, and God's knowledge in terms of reality.

Reference to God's omniscience does not address the basic question about the nature of reality. Yet once we assume that reality is constituted by scientific knowledge, it is tempting to extrapolate from this and envisage an ideal scientist, and per-haps even invoke God in that position. Needless to say, at that point human science has been left far behind. Unattainable knowledge may perhaps demand an unattainable vantage point, but it is hardly surprising that pragmatists and empiricists become impatient with the unattainable, as being metaphysical. Realists, on the other hand, will not be so concerned, because they have always accepted that reality and our views about it are logically independent. The impossibility of being able to reach a particu-lar vantage-point is not central to their concerns. Reality and how it appears from any perspective are distinct. A 'God's-eye view' is not needed to establish realism. Its opponents are entitled then to ask whether the concept of reality has not been emptied of all content, but that brings us back again to the question of the meaningfulness of metaphysics.

Ideal Science

Nicholas Rescher has tried to combine an emphasis on the active nature of the human mind in understanding and moulding reality, with a pragmatist emphasis on the concrete. Yet, he has also seen the importance of having a regulative ideal beckoning us on from where we are. We have already seen how the effort to combine realist and idealist insights in this way can create tensions in his work. It means that even when reality is beyond anyone's reach, it must at least be accessible in principle to understanding and experience. Like all pragmatists, he feels the urge to link what is real to what science might one day describe. It is not surprising, therefore, that he argues that a real object must be able to be experienced. He says that 'to be real is to be causally active – to be a part of the world's causal commerce' (1992, p. 282). He suggests that one can always suppose that there is a creature that can detect a certain sort of causal process. Therefore reality carries with it the idea of 'experientialibity' and Rescher advocates an 'experiential idealism'. Yet the ability to be experienced as Rescher describes it is a consequence of reality, and it is somewhat puzzling why this position should count as idealism. It is because something is causally active that Rescher thinks it can be detected by some creature somewhere, even if hypothetical. Unless the notion of causal activity is *already* explained in idealist terms, it cannot itself be incorporated in any idealist system. Yet if it is so explicable, causal activity cannot underpin experience. It must itself be accounted for in terms of experience.

For science to be able to explain everything, it must be possible to conceive of a perfect and complete science. Such science by definition must be a human creation, and our human limitations make this a forlorn prospect. Rescher himself states that 'science is imperfectible'. He continues:

> We are constrained to acknowledge that it is not present science, or even future or ultimate science but only ideal science that correctly describes reality – an ideal science that we should never in fact attain since it exists only in utopia and not in this mundane dispensation. (1992, p. 296)

He defends the notion of an ideal science as something that provides us with an aim. He says: 'The ideal of perfection, though unattainable, is nevertheless highly useful because it prevents us from sitting back in the complacent supposition that what we have is good enough' (p. 94). Science must aim for what it cannot reach, since that prevents scientists being content with what they know. We only have to look at the history of science to realize that wherever scientists have appeared, certain, subsequent discoveries have forced them to modify their views. Yet to suggest that this will be a never-ending process is to indicate that scientific progress aims at a goal that can never be reached. One is bound to ask whether such a goal is worth having. Rescher is in no doubt of its practical utility in guiding scientific research here and now. He considers it essential to retain some reference to truth in our understanding of the purpose of science, even if we couple it with a recognition of our limitations. He says: 'Without a reference to the truth, we would lose our hold on the teleology of aims and goals that define the very nature of the enterprise of scientific enquiry' (p. 57). To revert to an analogy used in chapter 2, a football match in which no goals are scored is very different from one where there are no goal posts. Indeed it would be hard to see the latter as a proper game at all.

Ideals may be useful but so may be illusions. A pragmatist might be satisfied with an ideal that had practical utility, but others might be more impatient and want to know what grounds, if any, it rested on. There is a significant difference between Rescher's position about truth as a regulative ideal that will never be reached and one which holds that truth can be grasped. It is often agreed amongst those who believe that science can progress that 'the ideal of comprehensive true theory . . . is essential to the scientific enterprise' (Kneale, 1967, p. 38). The problem is whether it is conceivable that we reach such a theory. Rescher's idea of truth is not so very different from Kant's *noumena*, things in themselves that are forever beyond our grasp. Others are more optimistic and with Kneale are willing to say: 'A scientist may someday formulate a comprehensive theory which is wholly true and therefore incapable of refutation.' Such a view explicitly dismisses any view of science which regards it as a succession of revolutions. The work of T. S. Kuhn has emphasized the fact of historical change in scientific beliefs (see Trigg, 1973 and 1985). One theory can be replaced by another which apparently populates the world with a new set of entities, as in the change from classical to quantum mechanics. This stress on revolutions merely focuses on change and suggests that the process will continue, without being able to explain the idea of scientific progress. We are then confronted with a series of worrying questions. Is there any point in the continuing flux of scientific belief? What is the purpose of change? Is there a direction that science is going in?

There are two traps in answering such questions. Rescher succumbed to the first. By making the goal unattainable, he runs the risk of producing the very scepticism which he wishes to avoid. The conclusion could easily be that if we can never really grasp the final truth of things, all our beliefs are equally mistaken. The second danger is to conclude that as the scientific beliefs of previous generations, even when firmly held, have been shown to be at best deficient, the chances are that *our* beliefs are equally misleading, and that future beliefs will fare no better. Both traps produce the conclusion that science is no guide to truth at all. Kneale's apparent optimism tries to avoid epistemological despair. What, though, is there to support it? The argument that science needs truth as a target is a fine 'transcendental' argument for continuing science. We have to have a strong sense of purpose, if we are to do science at all. Yet we do not have to be scientists, and if scientific practice has to be based or illusory targets, then perhaps we ought not to indulge in it. If, on the other hand we believe in the possibility of scientific progress because of the actual achievements and present success of science, we must acknowledge that science might even be able one day to produce a completely true theory of the world.

Many in fact do not see the history of science as merely a story of discontinuous change but take a rather more optimistic view. They see a steady increase in our ability to predict and control events in the world. They think that, despite the unfortunate by-products of many scientific discoveries, our ability, say, to put a man on the moon, proves something about our growing understanding of our environment. Science does, it may seem, get a grip on the real world, and there has, many would argue, been a piecemeal accumulation of knowledge.

Theories of Everything

Many physicists are now sufficiently emboldened to set their sights on producing a so-called 'Theory of Everything'. This 'will write all the laws of Nature into a single statement that reveals the inevitability of everything that was, is and is to come in the physical world (Barrow, 1991, p. vii)' The aim appears to be nothing less than to give an ultimate explanation of the Universe by giving the most basic and comprehensive version of the universal laws of nature. It is a paradox that the very rapidity with which modern physics develops should itself be a warning about the difficulty of obtaining complete and unrevisable theories. Yet it is in the midst of such development that physicists make their sweeping claims. Quantum field theories are no longer held to be the most basic theories of physics, and attention has switched to so-called 'string theories' which are mathematical in form.

The hope is that they will one day explain the properties of all the elementary particles. Such theories do not treat the most basic entities as points, and thus they avoid producing mathematical infinities. Instead they treat them as lines or strings. String theories can claim to be a Theory of Everything, on the grounds, John Barrow says, 'that they should contain within them the deep connection between the symmetries or laws of Nature, and the entities which those laws govern' (1991, p. 77). He continues concerning this would-be Theory of Everything: 'One day it is hoped that definite predictions of the masses of the elementary particles of Nature will be extracted from this theory and compared with observation.'

This urge to give a complete explanation seems to be what drives many physicists on, despite the constant changing of theories and the elusiveness of any end-point. Indeed it may partly explain such change. Is it fanciful to conceive of fundamental physics being completed? Those searching for a 'Theory of Everything' seem to be aiming for something remarkably like Peirce's scientific millennium. One problem is the mathematical character of physical theory. It may seem that a particular mathematical theory could be demonstrated to be both complete and unique, and if applied to the world, this might seem to be the unique and complete description of the world. This, though, highlights a problem about the role of mathematics in physics. Mathematics has always been thought to be the prime example of a body of analytic propositions or tautologies which are logically true but have no empirical content. The propositions of physics, on the other hand, can be tested through observation and experiment, and thus are synthetic. How can this philosophical gap between the analytic and the synthetic be bridged? A mathematical truth, however, complete, will tell us nothing about the world. It can never itself be a scientific theory. Even if we dismiss the distinction between analytic and synthetic as a dogma of empiricism, there remains a more metaphysical way of posing the same problem. Why should mathematics, the product of human minds, be so well adapted to be a description of the world, that the world apparently conforms to its demands? It is one thing for a physicist to observe the world and try to make sense through theory of all its apparent diversity. It is quite another to assume that the nature of reality can be discovered through mathematics alone.

The question is why any calculation, however consistent and complete it might be, should be applied to reality. Even the arguments that mathematical ability carries with it evolutionary advantage seems irrelevant to the calculations of higher mathematics. Humans have already survived a considerable time without access to such abstruse mathematical realms. Even if ability at such a level of mathematics was connected in some remote way with abilities important for survival and reproductive success, that is no ground for concluding that mathematics can now unlock the secrets of the universe. It may be important for biological survival that our minds are somehow attuned to the world that is our home. This is, however, a fact about our past evolutionary development, and can guarantee nothing about the accuracy of future mathematics. A facility with crossword puzzles could also be the out-working of an intellectual ability that has in other contexts proved its usefulness. It would be unnecessary to conclude that there was any analogue to crossword puzzles built into the scheme of things.

The claim that a Theory of Everything is possible in physics is not only the claim that physics will reach its goal and will never have to be revised. It is also the claim that such a theory will mirror reality mathematically, and that physical reality is at its deepest level mathematical. So far from being a Theory of Everything, the theory depends on an assumption that itself seems in need of some explanation. It apparently finds a rational structure in the world. If this is so, rationality is not a mere reflection of the human mind, even an essential one, as Kant would have us believe. The claim of completeness tells us that we have reached the final truth of things, as measured by the way things actually are. If rationality were merely a projection by the human mind on an unformed chaos, there is no reason why we should be impressed with any set of mathematical results. If mathematics is simply a construction of the human mind, it tells us more about ourselves than the world.

A Theory of Everything must, above all else, be asserting that the physical world is such that it can be understood though the exercise of reason. It somehow itself embodies rationality. Some, like Stephen Hawking, have appeared somewhat mischievously to raise a theological question at this point (1988, p. 175). Does such a rationality reflect 'the mind of God'? Certainly, without a theistic underpinning, the urge to obtain a Theory of Everything can result in a reliance on the power of human reason without any attempt to justify this. We cannot simply assume that the finite and fallible human mind can somehow see into the very nature of things. Saying that we cannot may be to deny the possibility of science. Yet saying that we can demands an explanation that must itself be beyond the bounds of science, since it will itself provide the grounding for science.

In 1980 Stephen Hawking looked forward to achieving the goals of theoretical physics by the end of the century. He said: 'By this I mean that we might have a complete, consistent and unified theory of all the physical interactions which would describe all possible observations' (1980). There still remains the question of what precisely has been explained, even if this could be achieved. Indeed, even if we have a complete theory, we may not be in a position to know that it is complete. One can always know something without knowing that one does. Further, it seems doubtful that we could ever be in a position to know what 'all possible observations' would be. Nevertheless, if everything *were* explained to the satisfaction of physicists, this is very far from giving us a complete knowledge. For instance, a Theory of Everything neglects the question of the emergence of complex levels of organization. Not everything can be explained in terms of the basic constituents of matter. To maintain that they had a Theory of *Everything,* physicists would have to be crude reductionists, insisting that everything had to be explained in physical terms. Society would have to be explicable in terms of the understanding of individual participants, their understanding in terms of neurophysiology, and neural systems eventually via biochemistry in the terms of fundamental physics. At each level holistic systems would have to be broken down into their constituent parts with the lower level explaining everything from the upper without remainder.

Science's claim to explain everything has, in fact, to face the varying claims of different sciences. We should not assume that science means only fundamental physics. The dogma of the unity of science all too often implies a reductionism which assumes that only physics is properly concerned with reality. It discounts the possibility of systems existing which cannot be explained simply by breaking them into parts. An organism cannot be described simply by an enumeration of the cells which constitute it. It is hardly surprising that different sciences correspond to what are allegedly different ontological levels. Biology deals with organisms, psychology the mind, and so on. Sociology's problem in establishing itself as a science can be traced partly to the fact that it is controversial where society forms a distinct level over and above individuals. Physicists, therefore, will find that their claims to have a complete theory will not go unchallenged from other branches of science.

Even if physics confines itself to explaining the basic nature of matter and to cosmological questions, it still seems to be involved in a conjuring trick when it claims to explain *everything.* Granted that reality is rational in a way that can be described mathematically, and that the relationship between mathematics and reality has been established, one is still left with the not inconsiderable question why there is anything rather than nothing. This is perhaps the most basic metaphysical question of all. No doubt it could be (and has been) dismissed as meaningless, but the sheer contingency of things, the fact that they do not have to be as they are, and indeed need not exist at all, is something that seems to demand attention. Even if it is demonstrated that the physical world has to be as it is, if it is to exist, there seems no way of showing that it *has to* exist. The necessary truth of a mathematical system cannot somehow be surreptitiously transferred to the physical world, so that it is necessarily true that it exists. Even God cannot be legislated into existence by the ontological argument. If He exists, it is no doubt true that His existence does not depend on anything else. It is not contingent. If He exists, He exists necessarily. This though does not make it necessarily true that He exists. Logic cannot conjure up His existence, and mathematics cannot do the same for the physical universe.

The Completeness of Physics

We might dispense with the idea of contingent initial conditions for the universe, but we still have not answered the question of why there is anything that behaves according to mathematical calculations published in the *Physical Review*. To paraphrase Wittgenstein, it is not a question of *how* the universe is, but *that* it is (1961, 6.44). His problem was that such a question strayed into the realm of the mystical, and only scientific questions could be put into language. He believed that 'when the answer cannot be put into words, neither can the question be put into words' (6.5). By definition, this apparently left science with an ability to answer every possible question, but Wittgenstein immediately remarks: 'We feel that even when *all possible* scientific questions have been answered, the problems of life remain completely untouched' (6.52). As we saw in chapter 1, he felt that all the really important issues lie beyond language. However, this is to concede too much to a scientific way of looking at things. The questions as to why anything at all exists and why there is any universe at all are not empty merely because of the difficulty of answering them. The idea that a question which cannot be answered by science cannot be properly asked is a dangerous one. Some questions may appear to be beyond the scope of science and yet be brought one day within its scope. The metaphysics of today, it could be argued, could be subsumed under the physics of to-morrow. This, though, merely acknowledges that science might *in principle* be able to answer every question, and in effect plays into the hands of the proponents of a Theory of Everything. What is at issue is whether in the end *everything* could be explained without any metaphysical residue. There seems to be no way in which physics could explain why there had to be a physical world. It might be shown that, given that there was one, it had to have a particular nature, but that is a different issue. It is, in fact, the one we have seen raised by the various anthropic arguments.

Stephen Hawking has been in the forefront of those who wish to claim that physics can explain everything. He has suggested that there are no singularities where the laws of science would break down 'and no edge of space-time at which one would have to appeal to God or some new law to set the boundary conditions for space'. He continues: 'The universe would be completely self-contained and not affected by anything outside itself. It would neither be created nor destroyed. It would just BE' (1988, p. 136).

A universe without a boundary, it is alleged, would have neither beginning nor end, and would thus have no need for a Creator. Yet just because, on Hawking's reasoning, a Creator is not needed to set up the initial conditions of the universe, his physical theory has not done anything to explain the existence of the universe. At times he recognizes this and points out that if there is only one possible unified theory, it is just a set of mathematical rules and equations. There still has to be a universe for them to describe. He asks:

Why does the universe go to all the bother of existing? Is the unified theory so compelling that it brings about its own existence?

Or does it need a creator, and, if so, does he have any other effect on the universe? (1988, p. 174)

No mathematical theory 'could bring about its own existence'. Theories can exist, but still not be *about* anything. What Hawking is concerned with is whether a theory corresponds to the real world. No theory, however logically compelling, can conjure a world into being. If we start with the laws of physics, expressed mathematically, all we are confronted with is the human mind. It bears repeating that logical necessity cannot produce ontological necessity. Even if it could, a Theory of Everything operates at such a high level of abstraction that it cannot explain a simple organism like an amoeba, let alone the special circumstances of human life.

One tempting argument against a Theory of Everything would be to deploy Gödel's theorem against it. This showed that if arithmetic is consistent, its consistency cannot be established by any mode of reasoning that can be formulated within its domain (Nagel and Newman, 1959). No theory can show itself to be consistent. This alludes to the deep and pervasive problem concerning the difficulties of self-reference and reflexivity. The whole of reasoning is affected by it, and very often a theory or principle which ignores it can destroy itself. At a simple level, the problem is illustrated by the ancient example of the Cretan who claimed that all Cretans were liars. He must have been lying if he was speaking the truth. At a more global level there is a problem of the person who claims there is no such thing as truth, but appears to be asserting this as true. As we have already seen, this is a constant temptation for thinkers. According to Plato, Protagoras fell into this trap, but at the present day, many are inclined to glory in such 'reflexivity'. Yet Gödel's theorem shows the need to step outside a system, even in mathematics, in order to enable us to justify it. Every formal system contains formulae which can be seen to be true, but which cannot be proved within the system itself. Thus J. R. Lucas points out that all mathematical systems are 'incomplete' (1970, p. 128). He claims: 'Every serious mathematical system has some statements in it which cannot be proved in it, but which we should have to agree were true.' The principle can perhaps be extended to other modes of reasoning, besides the mathematical, but it could be particularly relevant to physicists who use mathematics to explain everything.

Does Gödel's theorem show the task of total explanation is impossible? It certainly suggests that there are difficulties in trying to explain a whole system, when we ourselves are members of it. A Theory of Everything could not explain its own existence without producing an infinite regress. The more specific mathematical point is not such a threat. Unless the Theory used the whole of mathematics, there is no reason why its consistency could not be shown by mathematics which is not itself deployed in the theory. A particular mathematical calculation can always rely on other parts of mathematics.

Even wider claims have been made on the basics of Gödel's proof, and substantial conclusions have been drawn about human rational capabilities. For example, it has been suggested that the theory shows that the calculational procedures, or 'algorithms', according to which computers operate, can never on their own be sufficient to ascertain truth No system, however apparently self-contained, can exist in a vacuum. It has to depend in the end for its force on the simple acknowledgement of something as true. Roger Penrose, the mathematician, has argued that decisions about the validity of an algorithm is not itself an algorithmic process. It depends instead on somehow stepping outside the mechanical procedure and simply *seeing* that something must be the case. Indeed he thinks that this is the essence of mathematical understanding, and says: 'We must "see" the truth of a mathematical argument to be convinced of its validity' (1990, pp. 554, 541). He enlarges his claim by going on to link such seeing with the nature of human consciousness. He then uses Gödel's theorem to show the indispensability of our ability to reason and to grasp truth. We have to be able to break away from the mere following of procedures, and to have the capability of recognizing their validity or invalidity. Reason and truth are then tied in with the ability of human consciousness to rise above the demands of any system, whether it is mathematical or not, and to grasp what is the case. It would seem that if humans were not able to do this, we could never be able to talk of truth. Human rationality, which must depend on the ability to separate truth from falsehood, rests ultimately on it.

Penrose sees in all this the nature of human consciousness, and says that the hallmark of consciousness is 'the ability to divine (or "intuit") truth from falsity (and beauty from ugliness!) in appropriate circumstances' (1990, p. 533). This does, of course, involve a rather restricted notion of consciousness, which connects it more to rationality then to the simple ability to feel. Such appeals to intuition are, in addition, somewhat dangerous in that they often ignore human fallibility and the inconvenient fact that different people can have many conflicting 'intuitions'. This is particularly the case in areas such as those of aesthetics which Penrose mentions. Too great a reliance on 'self-evident' truth can block rational argument and discussion as well as provide a way of grounding it. Appeals to intuition can easily become the enemy of human reason.

It is, however, perhaps not surprising that mathematicians should emphasize the importance of self-evident truth, since philosophers have always recognized that mathematics, above all other areas of human thought, is the most likely to be the repository of such truth. In that discipline at least, the human ability to recognize truth can offer a way of breaking out of the strait-jacket of algorithmic reasoning. Our minds are more than elaborate computers, which cannot on their own recognize the truth underlying the procedures they are programmed to follow. An appeal to intuition also offers the attraction of blocking an infinite regress, when each step is justified in terms of another step. Instead we can come to a halt in the perception of truth. Top mathematicians, like Penrose, can claim with some plausibility to have such experiences, although a sign of their validity must be the ability to persuade others of the truth of what they have 'seen'. The rest of us can perhaps share in this when we come to recognise how some mathematical calculation, however simple, *has* to be true. The recognition of necessary truth is itself an important human capacity, which underpins much of our rationality. Not all human reasoning is like this, however, since, apart from other considerations, necessary truth is only a small part of what may be true. Intuition will never suffice when it comes to grappling with the changing circumstances of the contingent world of which we are a part.

Chaos

The idea that science can explain everything has been aided by the fact that the word 'explanation' has tended to be monopolized by the physical sciences. As a result the notion that explanation must be scientific seems almost tautologous. It is a commonplace that science was liberated by the overthrow of scholastic philosophy with its insistence on the role of final causes or purposes in the scheme of things. Aristotle's four types of causation have been whittled down so that only his 'efficient' cause has been regarded as an insistence of genuine causation. It follows that unless science can formulate a law showing the operation of cause and effect in a particular case, it has seemed that no explanation can be forthcoming. Indeed it has often been assumed, particularly under the influence of classical mechanics, that science is deterministic and that any event can be predicted, given the laws of nature and a knowledge of past events. Indeed this process has appeared to be the very summit of rational understanding. It has certainly proved advantageous to adopt determinism as part of the methodology of science. If one does not look for causes, one assuredly will not find them. The assumption that reality is of its nature deterministic is more problematic.

The indeterminacy intrinsic to quantum mechanics suggests deep obstacles in nature at the subatomic level to any simple notion of causal explanation (see Trigg, 1989, chapter 6). This itself sets limits to the possibilities of prediction. Recent advances in the field of so-called chaos theory point to the same conclusion. Chaotic systems are extremely sensitive to initial conditions. Instances include the turbulent flow of liquids or the dripping of a tap. The weather is the most obvious example and the 'butterfly effect' is often remarked. Minute initial differences can be amplified so that a movement as slight as the flap of a butterfly's wings on one side of the world can eventually result in a major storm on the other. The point can be taken further, in that the behaviour of a single electron, itself impossible to predict, may in the long run produce a major calamity.

So far from the world being a mechanism with predictable and determinate behaviour, it is now clear that it is a much more open and flexible system. It is more like a cloud than a clock (cf. Popper, 1982, p. 6). This perhaps begs the question as to whether the problem with clouds is not our lack of information rather an intrinsic unpredictability. It would be alleged that we might predict the storm if only we could know about the butterfly. The difficulty, it might seem, is not theoretical at all, but the practical one of knowing every infinitesimal movement by every creature. Therefore it is not a simple question about the breakdown of causal explanation, but a problem arising from the limitations of our knowledge. Popper's claim is that 'scientific' determinism requires the ability to predict every event with any desired degree of precision, providing we are given *sufficiently* precise initial conditions. Chaos theory may suggest that we are never able to grasp the initial conditions of much of the behaviour of the physical world with the necessary degree of precision. Even discounting the relevance of quantum indeterminacy for a moment, we are confronted with the problem whether chaos theory demonstrates a limitation on the possibility of our knowledge about the world, or whether it somehow goes further and shows us a deep truth about the workings of nature.

The question of an underlying indeterminacy built into the very scheme of things as opposed to the 'mere' impossibility of scientific prediction is an important one. It tends to be blurred by those who see no difference between metaphysics and epistemology and who tie the limits of reality to the limits of human knowledge. Nevertheless it is obvious that the butterfly effect does show the powerlessness of human science to make predictions beyond a certain point. Weather forecasting cannot be accurate more than about four days in advance. Long range forecasting extrapolates future trends on the basis of past ones. Forecasting is powerless when the weather behaves in an unusual way, as it is at times quite capable of doing. Massive storms can blow up even while forecasters are denying the possibility. The initial conditions for such events cannot be specified with the requisite precision. Indeed it is hard enough to describe what the weather is actually doing at any given moment with absolute accuracy.

The mathematical precision required for prediction must, in fact, be infinite. For instance, precision to ten decimal places will be accurate for short-term predictions, but variations beyond that will be amplified and may eventually produce cataclysmic effects. What seems an irrelevant and infinitesimal difference in the short-term, and one which is undetectable, can be responsible for the widest variations possible. If one can imagine a series of billiard balls knocking into each other, a minute variation in the way the first ball is touched may not have a noticeable effect on the way it hits the second. It will however become magnified as the second hits the third, and so on. Within a very few balls it will become impossible to predict the way a ball will go. This is not just a practical limitation. However accurate the prediction is, it will always be possible to imagine a more accurate one, perhaps to a further decimal place. Chaos theory shows that this does matter, and that what is at issue is our ability to go on making ever more accurate predictions infinitely. This is in principle impossible, and we seem to have reached a major limitation on the scope of scientific explanation even within the terms it sets itself.

Another related example of the limitations encountered by science is provided by measurement, which has always been thought to be crucial for the conduct of experimental science. Some measurements, such as of the simultaneous position and momentum of an electron, are impossible to make in principle (see Trigg, 1989, chapter 6). Yet the very idea of an absolutely accurate measurement at all is now seen to be an impossibility. It is the discovery of chaos theory that this actually matters. Mathematicians now deal with so-called 'fractals', geometric objects which continue to show detailed, and indeed the same, structure on a wide range of scales. One mathematician writes that 'an ideal mathematical fractal has structure on an infinite range of scales' (Stewart, 1989, p. 217). Snowflakes provide an example of naturally occurring fractals. Coastlines provide a similar example, in that it is impossible to obtain an absolutely accurate measurement of the length of the coast of Britain. We can be accurate enough for most of our purposes, but even if we have measured the circumference of every bay, it would be unlikely that we had measured every indentation in every bay, or every twist of every rock. For each accurate measurement, we can easily imagine one that is more accurate and so on to infinity.

Chaos and fractals deal with the 'structure of irregularity' (Stewart, 1989, p. 216) and each illustrate the problems science faces when confronted with a world that fails to conform to the apparent regularity of a clockwork mechanism. Scientists from the seventeenth century onwards have perhaps taken the inherent order and regularity of the physical world too much for granted. It has to be presupposed by science, rather then proved by it, since without it all science would be impossible. The continued success and progress of the physical sciences has blinded many to the fact that science cannot be taken at face value. It has been too easy to take actual physical science as a starting-point, and in Kantian style enquire into the presuppositions of its possibility. To be successful, it was argued, science must assume a deterministic world where cause mechanically produces effect. Science is self-evidently successful. Therefore it seems that the world has to be deterministic. Yet during this century, first in quantum mechanics and then in other areas, it has become apparent that science, even within its own terms, has come against barriers that it cannot set aside. Even the idea of a repeatable experiment, which has been one of the cornerstones of science, must be questioned. Any situation in which initial conditions matter, such as the collision of atoms in a gas, is one which can never be repeated *exactly*. The problem is that as science turns from macroscopic to microscopic entities, it encounters new difficulties. It may explain the turbulent behaviour of a fluid as a whole, but it cannot deal with particular particles. Yet even whole systems, as in the case of the weather, do not behave in regular ways. One could never get two *exactly* similar weather patterns.

Even within science, there are limits to the possibility of prediction, the accuracy of measurement, and the repeatability of experiments. As a result, there must be limits to the very ideas of verification and falsification which have played such a large part in the philosophy of science during this century. What counts as verification or falsification will depend, amongst other things, on the standard of precision required. Whether one situation is the same as another depends on how closely it has to be specified. To take the case of weather forecasting again, more sophisticated and more widespread equipment can help. Satellites have enabled significant advances to be made. However, the point is that after a certain point it is not just a matter of practical impossibility to describe every subtle and simultaneous variation in weather. It is in principle impossible beyond a certain degree of precision. One can always imagine a more precise description, and that means that one can never exhaust the complexity with which we are confronted.

It may seem surprising, in the light of all this, that science is possible at all. What seems to be at issue is not so much how we explain the obvious success of science, but whether science has the right to claim any success. So far from explaining everything, physics has to recognize its limitations in the face of infinite complexity. For science to be possible, the world has to be ordered and structured in some way, but it does not follow that a scientific description of the world must be complete. If the world contains a subtle interplay of order and chance, of determinism and randomness, even of causes and freedom, science will have plenty of scope in uncovering the nature of things. What it will not be able to do, even in principle, is to put us in the position of being able to predict infallibly, or control absolutely, the world of which we are a part. It is a paradox that the mathematics of chaos simultaneously increases our understanding of physical reality and shows us more clearly what we can never fully know.

Research into the character of natural irregularity has demonstrated that what science can do depends on the nature of the reality it confronts. Science can produce a complete set of deterministic laws only if physical reality is itself deterministic. Science must mirror reality, discovering what it is like. Yet the reality under investigation did not have to be as it is. The sheer contingency of the world is a fundamental datum of science, and is the reason why physical science has ultimately depended on sense experience. One has to find out the way things happen to be. As we have seen, it is not necessary for the world to exist at all, let alone have the characteristics it happens to have. The necessary truth of mathematics cannot be assumed to reflect the contingent truths of the physical world. The Greeks always assumed that the dictates of reason would demonstrate the truth about the physical world. They excelled in geometry but did not see any pressing need for observations of the actual universe to confirm that it behaved as they thought it ought to. They were unable to formulate anything like Newton's laws of motion. Aristotle assumed that the rate at which bodies fall to the earth is determined by their weight (1939, 274a). Yet no Greek bothered to perform the experiment which tradition ascribed to Galileo, who was said to have dropped objects from the top of the leaning tower of Pisa. This curious omission was not the result of any lack of curiosity or carelessness. If the world behaves according to the rules of reason, empirical investigation is not needed to confirm what must be the case. No one indulges in fieldwork to interview bachelors to see if they are unmarried, or to manufacture triangles of some material to see how their angles add up. The realization of the contingency of the world lies at the heart of all modern science. We have to test and confirm to find out what the world is actually like, because it does not need to be as it is.

The modern recognition of contingency and the ensuing emphasis on the importance of intersubjective experience as a source of knowledge produces a problem which is the reverse of the one confronting the Greeks. Instead of discounting experience, we face the temptation of undervaluing the place of reason in science. The question is that if the world is simply as we find it to be, reason seems to be little more than an instrument in our investigations. The limits of science are not then set *a priori* but are simply the barriers we find in place. The limits appear to be contingent derived from the nature of the world as it happens to be. This seems to be the lesson of twentieth-century physics, which itself has uncovered the complexity of things. History has also shown the folly of those who have been too confident in predicting limits to the possibilities of science.

Are the limits of science only contingent ones after all? Certainly, the world could have been less complex. Yet the very notion of contingency is a philosophical one. Modern science became possible because scientists consciously came to the understanding that the world is contingent. The very practice of science, with its emphasis on empirical investigation, depended on a prior philosophical belief, the importance of which is shown by the gulf that exists between Greek science and modern science. Science is partly limited by its need of a philosophical basis on which to rest its understanding of itself and its purpose. There are, as well, rational limits to the possibility of a complete science. The simple fact that scientists are part of the world they are describing has enormous repercussions. There are intrinsic limits to the possibility of scientific prediction. For one thing, we cannot predict our future discoveries before we have made them. As Popper has claimed, 'Knowledge . . . cannot foreshadow its own future conquests. I cannot forecast future inventions in any detail, without myself being the inventor' (1982, p. 109). This is just one instance of the limits to the extent to which human reason, as expressed in science, can explain itself. Just as we cannot anticipate genuine creativity, the scientific project seems to carry with it an inevitable incompleteness. For one thing, it can never fully explain itself. Popper illustrates this by referring to the man 'who draws a map of his room, including in his map the map which he is drawing', and concludes that 'his task defies completion, for he has to take account, within his map, of his latest entry'. This type of self-referential situation always poses the threat of an infinite regress. For this reason alone, humans can never hope to attain to a complete knowledge of everything, including the fact that they possess a complete knowledge. Limits placed on human understanding inevitably must cast their shadow over science itself.

9
The Legitimation of Science

The Purpose of Science

The physical sciences have consistently taken their apparent success for granted, and have all too often scorned the need for justification. The empiricism of Hume, for example, is notable for its inability to ground science in reality. Nowhere is this more apparent than with his problems about induction. The uniformity and regularity of natural processes have always to be presupposed by science. It cannot be empirically discovered, since it is notorious that all we can ever discover is that up to now things have behaved in a regular fashion. Even the existence of reality is a problem for science, as the perennial issues concerning realism indicate. Again, we cannot prove its existence through the practice of science, since that begs the question, and further assumes that reality, or part of it, is scientifically accessible.

These are profound philosophical questions, and witness to a recurring desire to ground science in something beyond itself. In particular, many have wanted to appeal to an apparent rationality and order built into the very nature of things. We have seen in the course of this book how contemporary philosophy often follows empiricism and pragmatism in rejecting the possibility of such metaphysical appeals to reality. The results can be the kind of explicit and debilitating relativism we have encountered, or at least a superficial resignation to playing the scientific game without really knowing why. Yet a hunger for justification remains, even amongst working scientists. They may not want to feel that they are merely projecting the prejudices of their own society, or worse, their own small group of colleagues. Contemporary rejection of the possibility of reason will not impress any scientists who follow Einstein in the awe he felt at recognizing an inherent comprehensibility in the physical world. In a letter to Solovine, he dealt with his friend's objection of the use of the word 'religious' in this connection. Einstein's response was that he had no better word 'for this trust in the rational character of reality and in its being accessible, at least to some extent, to human reason' (quoted in Fine, 1986, p. 110). Einstein recognized, as he said, that priests might make capital out of this, but was afraid that otherwise science would degenerate into 'senseless empiricism'. That indeed has all too often appeared to be the choice. On the one hand lay a reversion to the Middle Ages and the stifling authority of the Church. On the other hand there was an empiricism that dealt with the amassing of 'facts' with nothing to bind them together, or to give them any justification. The multiplication of collective experience may be impressive but it can give no guidance as to the nature of things or their future behaviour without some reason for generalizing from them. We need theories to make sense of isolated experiences and such theories will inevitably go beyond momentary insights and make assertions about the nature of the physical world. They will inevitably be underdetermined by our experience, since, in any case, experience can only ever give us a partial glimpse of a persisting world. Yet, although we may only be able to approach physical reality through our theories, we still need some assurance that there is something for our theories to be about. They themselves have to be grounded in a prior understanding of the world.

This urge to give science a grounding is still resisted by many philosophers of science. Arthur Fine, for example, reiterates what should now seem a familiar theme by saying that the only reason we can have for believing in molecules, atoms and quarks, are those embedded 'in the various overlapping and ever-open practices that constitute the judgement of these claims by the community of concerned scientists' (Fine, 1986, p. 171). We have to trust the judgement of scientists, because that is all there is. Fine acknowledges that his attempt to ground scientific belief in 'reasonable practice' has 'none of realism's splendor'. It does not, he says, see in actual scientific endeavour, 'a reaching out to the hidden details of an exquisite and elaborate world structure'. We have to let science speak for itself and 'to trust in our native ability to get the message without having to rely on metaphysical or epistemological hearing aids' (p. 150). What is the message? The right of science to speak meaningfully must depend on its alleged relationship with physical reality. Yet, as we have seen, science cannot justify itself in its own terms. Its traditional enmity with metaphysics or religion seems to preclude its turning elsewhere for justification. The result is that those with a scientific outlook will often be contemptuous of the idea that one needs a reason for doing science or trusting scientists.

Fine's own approach epitomizes a common reaction to a search for a grounding for science. We might ask what is its aim or purpose. Since he has declared that 'realism is dead' he cannot talk simply of discovering truth (1986, p. 112). Instead he points out the similarity of the question to the one about the purpose of life. He snidely suggests that 'as we grow up, I think we learn that such questions really do not require an answer.' That may not prevent people asking them, but Fine says firmly of such questions: 'They call for an empathetic analysis to get at the cognitive (and temperamental) sources of the question, and then a program of therapy to help change all that' (p. 148).

Anyone arguing that science needs a purpose will not be given a fully satisfactory response. Instead the problem will be hurriedly swept out of sight, and the questioner blamed for some deficiency. Realists, similarly, will not even be allowed to make their claims. According to Fine, Einstein's realist language should not be taken at face value at all, since it certainly does not express any beliefs about reality. He says: 'It is rather the dues that Einstein felt worth paying for his passionate commitment to science, and for the meaning that scientific work gave to his life' (1986, p. 111).

The argument is that since realism cannot be a metaphysical thesis, it must be understood in some other way, as the expression of scientific motivation. Any language referring to reality is thus simply a sign of one's intense commitment to the scientific enterprise. Realism has been turned on its head. Far from being a justification or explanation, it is the outcome of attitudes already held by scientists. This method of argument follows a characteristic pattern. When science has not been able to explain something, for instance the validity of moral judgements, because it is not susceptible to public checking, the tendency, as we have seen, has always been to consign it to the private sphere. It is regarded as a subjective, personal matter and not an objective, public one. Fine adopts the same posture with regard to the problems of grounding science. Anything that cannot be settled scientifically is made to appear a matter of individual motivation or an expression of an attitude. Science itself therefore is not to be regarded as the discovery of reality. Many scientists, and Einstein in particular, can be seen as thinking this as a way of giving their activities meaning. Meaning in life and in science is arbitrarily conferred by participants. It is a matter of psychology not of metaphysics.

The proponents of science have often resorted to this strategy. One advantage is that subjects that seemed beyond the reach of science are actually brought back within its range. If attitudes can be portrayed as devoid of any designs on truth, but as mere matters of fact about individuals, then explanation can be sought why anyone should possess such an outlook. Sociological, psychological and even psychoanalytic accounts can be produced. Indeed the fact that so many disciplines are able to advance explanations with an apparent scientific pedigree is itself somewhat worrying. Science does not talk with one voice, and although some may hanker after a reduction into one basic physical explanation, there are many scientists who would vehemently oppose this. The battles between sociologists and sociobiologists have been exceptionally fierce (see Trigg, 1982).

'Science' does not provide one clear and uncontested explanation. Its methods often provide a licence for different explanations at varying levels, which often are unlikely to be all equally right. This is particularly so if they aim to provide more than a partial glimpse of the constraints on us. If, for example, our beliefs are wholly formed by society, much further argument is needed to demonstrate that psychoanalysis is *also* correct in attributing them to personal histories. Arguments between scientists often seem more like disputes over the possession of territory than a rational appeal to truth. Yet questions about truth will not go away even if we restrict matters to what can be settled within science. Sometimes attempts are made to reduce the claims of one apparent science to another, for instance psychology to neurophysiology, or another ploy is to deny that a particular discipline is a science at all. Sociobiology and psychoanalysis have been attacked in this way, as 'pseudo-sciences'. Scientific method alone is not sufficient to arbitrate, as what is often at issue is the kind of explanation required. Should one offer one at the level of social structure, say, or of genes? Truth cannot be easily identified with what scientists say, or might one day say, since this begs the question as to who is to be counted as a proper scientist.

A willingness to use empirical methods does not guarantee truth. This is not just because of the fallibility of scientists, but is also linked with questions about the aims of particular sciences. Saying that science should discover reality is itself not a great help in giving guidance. It puts some constraints on scientific conduct, ruling out wishful thinking and deliberate deceit. Each scientist, however, has to approach a segment of reality, and decisions have to be made about what is being looked for and at what level. Scientists have to decide whether they are physicists or biochemists or whatever. Each scientific discipline depends on a theory about the character of reality, and this shows the importance of what Kuhn called scientific 'paradigms'. Yet, despite his strictures on the incommensurability of theories, it is obvious that paradigms can and do conflict. This is true of competing theories within a discipline and also of competing disciplines. Questions of truth cannot be shirked. We still have to decide which aims of which science should be adopted. Should biology, for example, be concerned with the holistic study of organisms, or the replication of genes? Even within one science the question of aims is intimately linked with what is considered real.

It may be accepted that there are deep disagreements within science as well as beyond it. Yet, it will be said, this still stems from the sociological or psychological roots of such disputes. Science can thus explain why the dispute has arisen, and it might still be claimed that disagreements within science can be resolved because there is a recognized method for doing so. Yet a circularity of reasoning lurks behind this assumption. We are told that something is scientific if there are clear methods available for resolving disputes in a way that arguments outside science cannot be setttled. Yet simultaneously we are being informed that apparently intractable disputes within science, which scientific method seems powerless to resolve, must be soluble. The reason given is that they are scientific disputes. It just has to be accepted that agreement is not necessarily more widespread within science than within any other part of human reasoning. Indeed fierce disagreement between scientists, even over very fundamental issues, is an important ingredient in the testing and proving of ideas. Often different understandings of the nature of the world are at stake. Once theory is involved, and we begin to transcend, for good scientific reasons, the ordinary empirical world, scientific agreement about what is or is not relevant becomes much harder to secure. That is one reason why there is a lingering reluctance on the part of some empiricists to allow that theoretical entities are more than theoretical constructs. The urge to keep science based in the public world which we all share, and supposedly agree about, is still strong.

The Problem of Consciousness

The desire of scientists to be objective, in the sense of being able to secure maximum agreement, leads to a concentration on what is publicly accessible. This does not merely show itself in reducing non-empirical claims to truth to the status of attitudes, but also becomes prominent in other areas. As we shall see, there is a great debate about the nature of human consciousness, but similar problems occur in the science of ethology, which examines the development of animal behaviour from Darwinian presuppositions.

The tendency is for ethologists to eschew so-called 'subjective' language about awareness, and to concentrate on animal behaviour. Their grounds for this are that even if some animals do possess definite mental states, the latter are only of significance for a Darwinian if they emerge in behaviour. Fitness is affected, it is alleged, by variation in behaviour and not mental experience. This approach deftly avoids the need for apparently fruitless controversy over something which the resources of science are inadequate for settling. It begins as a methodological issue that science will not deal with what is apparently beyond its reach. Yet it is all too easy for this to drift into an ontological claim about the absence of any animal consciousness, or awareness, at least at any significant level. What appears irrelevant to science can easily be ruled out.

The example of ethology, however, suggests that a narrowly scientistic approach can actually debilitate science itself. If many animals do possess some form of subjective awareness, science should not ignore this, since the question arises why it exists. Even from a Darwinian point of view, it is cavalier to dismiss this question. It may seem as if two organisms with the same inputs and outputs, but one with consciousness and the other without, will be equally susceptible of scientific understanding. All that matters, it seems, is the way their behaviour promotes or limits biological advantage. Yet it does not follow from the fact that science finds questions about animal consciousness difficult to deal with that consciousness may not be a significant factor in evolution. The ability of an animal to be aware may help it to be more flexible in dealing with the environment. It can remember, for example, and recognize sources of danger. The feeling of pain may be particularly useful, even in animals, as a warning signal. It is clearly open to discussion which creatures can feel pain, but the ability to do so must, if it exists, have itself evolved, and should be a proper study for ethology. The question why it has evolved should be asked and answered. Yet many ethologists feel that delving into such 'subjective' issues as what an animal feels is itself 'unscientific'.

Consciousness has always been the ghost haunting discussions about animal and human behaviour. Some have always been tempted to define it out of existence, precisely because it appears to go against the grain of 'objective' science. Yet if it exists it cannot be an unimportant epiphenomenon, but rather a central fact about some animals and even more so about humans. It would be extraordinary from a Darwinian point of view, if awareness of one's surroundings, let alone other characteristics of consciousness, played no part in biological fitness. Dismissing it as an uninteresting consequence of increasing neural complexity is the result of an eagerness to keep science and its objects in the public domain. The objectivity of science has in fact little to do with objective truth. Subjective experiences accessible only to an individual may still be real and part of the objective world. They are not however part of the public world. Science is always in danger of being too concerned with what is publicly accessible and too dismissive of what is not.

Many philosophers take it as obvious that to be real a property, or whatever, must fit into a scientific account of the world. Arguments rage in the philosophy of psychology about the status of so-called 'qualia', the distinctive characteristics of subjective experience, whether the colour of a sunset, the sound of a waterfall, or the feeling of pain. We have already seen how some are ready to eliminate all references to mental phenomena from our language, and are contemptuous of 'folk psychology'. Daniel Dennett has written that 'the challenge is to construct a theory of mental events, using the data that scientific method permits' (1991, p. 71). He acknowledges that this will have to be constructed from a third person point of view, and it would seem inevitable that in such a theory 'qualia' will disappear. The inputs and outputs for an organism will be emphasized and various forms of internal processing will be flattened out, so that no distinction need be made between what is conscious and what is not. To put it another way, consciousness need no longer be understood as being some inaccessible, private experience. Dennett says that 'there is no reality of conscious experience independent of the effects of various vehicles of content on subsequent action'. He denies that there is a 'Cartesian Theatre', 'a place where "it all comes together" and consciousness happens' (p. 39). One of his prime reasons is that the image is 'scientifically unmotivated' (p. 132).

Since he believes that reality must be scientifically accessible, he alleges that 'postulating special inner qualities that are not only private . . . but also unconfirmable and uninvestigable is just obscurantism'.

The Cartesian approach to knowledge was to begin with ourselves and decide what we could really be sure of, before trying on that basis to build up a picture of the world. The scientific approach on the other hand starts with science, and has no hesitation challenging the existence of what each of us might feel most sure of, namely how things look and feel. The intrinsic quality of pain, for example, as distinguished from that of other sensations, just disappears under the scrutiny of the scientific investigator (see Trigg, 1970, for a defence of this notion that pain has an intrinsic quality). Yet for science to conclude that there is no such state as my feeling a pain is exceedingly dubious. It is verificationism returning to haunt us. Dennett denies that there could be 'qualia', but then rather puzzlingly acknowledges that there seem to be. This does not mean that an experience has intrinsic properties. Dennett rejects the view that anything 'really' seems to be a quality of experience. His point is that even if something seems to be the case, this does not mean that I have an experience of it seeming to be so. Dennett claims: 'There is no such phenomenon as really seeming – over and above the phenomenon of judging in one way or another that something is the case' (1991, p. 364). Thus it appears that for him a quality of an experience is merely a judgement, and it can be alleged that robots could be designed to make judgements, even if in a somewhat derivative sense. Certainly judgement can be anchored in the public world by being expressed in subsequent behaviour. He writes: 'Only a theory that explained conscious events in terms of unconscious events could explain consciousness at all. If your model of how pain is a product of brain activity still has a box in it labelled "pain", you haven't yet begun to explain what pain is' (p. 456).

Yet it could be claimed that anyone who does not know what pain feels like has failed to get a grip on its most important feature. Those who are congenitally insensitive to pain, for instance, are like those who are colour-blind (see Trigg, 1970, chapter 9). They are missing an important aspect of human experience, which cannot be captured properly by reference to neurophysiology. Even if someone's insensitivity were the result of a physical disorder, the latter could only be of significance *because* of the former. It would indeed not be logically imposs-ible, and possibly not physically impossible, for someone with a similar neural malfunction to be able to feel pain.

There is a sense in which a scientific theory of consciousness may be possible. We may for instance be able to find out what level of neural complexity is necessary to produce consciousness. What science cannot do is explain away the distinctive features of consciousness while pretending it has not done so. Conscious-ness, and the further ability to be self-conscious and reflect about one's own states, are eliminated by a scientific programme at the cost of bringing into question the very status of science. Dennett holds: 'There may be motives for thinking that consciousness cannot be explained, but . . . there are good reasons for thinking that it can' (1991, p. 455). The crux is whether explaining con-sciousness in the impersonal and 'objective' terms of science may not itself actually undermine the idea of there being good rea-sons for anything. The very notion of good and bad reasons goes with an ability to make a rational assessment, and decide on the evidence whether to accept a certain position or not.

It is hardly surprising, given his attack on the idea of a 'Cartesian Theatre', that Dennett is hostile to any idea of there being a self, which is the subject of experience. He says: 'Our tales are spun, but for the most part we don't spin them: they spin us. Our human consciousness, and our narrative selfhood, is their product, not their source' (1991, p. 418). As a result, there is no real subject, able to make rational decisions. Science has not only undermined the reality of private experiences but has apparently destroyed the subject of that experience. That need not be so very surprising, since any empirical approach will find it harsh to discover an experience of the self. By definition it is always the self which is having the experience. Dennett, though, is not the kind of empiricist who builds an edifice of knowledge on the foundation of private experience. For him, science sets the stand-ard of acceptability, and science will never be able to produce a self, as opposed to a body or a brain, in the public world. Yet scientists who make decisions about what evidence to admit are themselves rational subjects judging truth. If they are not, the force of their conclusions is lost. Dennett wants to talk of good reasons for thinking something true, and yet challenges the whole basis for talk of any subject having reasons for belief.

The Destruction of the Subject

The conviction that there is no place, even in the brain, where 'it all comes together' calls into question the possibility of rational judgement. The brain may be very good at all kinds of techniques of parallel processing, but the idea of reasoning vanishes without some central control acting as a unifying factor and able to reflect in a self-conscious way on all the information available. All we would be left with would be the sophisticated interpretation of stimuli in many simultaneous ways. There would be no subject pulling it all together or deciding what to believe. Consciousness is the necessary pre-condition for rational reflection, giving me the chance to make sense of my experiences. Without a centre of consciousness 'I' do not exist, and therefore *I* cannot make any rational judgements. Even the dividing of 'qualia' into judgements about what is the case leaves unresolved the question of who or what is judging. Similarly the claim that there seem to be qualia leaves unasked the question to whom this illusion occurs. Our language may be involved with a mistaken metaphysics, but the destruction, or perhaps deconstruction, of the subject would put in question everything we wish to say, since it is no longer clear who 'we' might be.

These objections might not appear relevant to philosophers who believe that it is possible to give a scientific account of everything, including the self. They may dissolve matters into a myriad brain processes but could still maintain that an account of brain function *is* an account of the self. There could be nothing left over and above this, given a physicalist approach. Dennett himself is even opposed to so-called 'Cartesian materialists' who hanker after some part of the brain as a central controller. He believes that positing a subject, whether material or immaterial, 'where it all comes together' is just to stop trying to give an account of the subject. If his explanation in turn still calls on a subject co-ordinating the brain, the argument is that this itself generates an infinite regress. We explain the subject in terms of diffuse neural activity only to be confronted again with a subject in charge of that activity. Dennett calls this a 'homunculus', a little man inside the brain. If we embark on this line, our explanation appears to have generated a need for a precisely similar form of explanation again. We have achieved nothing since we now have to explain the functioning of the subject which is in charge of the neural activity. That subject, once explained, will no doubt be seen as being guided by a further subject, and so on.

This is a powerful argument, but it illustrates the inadequacy of any attempt to explain the self in terms acceptable to science. Dennett gives up the attempt by denying that there is anything left to explain. For example, language generation itself is simply a matter of 'opportunistic, parallel evolutionary processes' which he dubs 'Pandemonium' (1991, p. 242). Thus, who 'I' am and what 'I' say and mean, can be fully accounted for in terms of multiple physical processes, with no 'Central Meaner'. Indeed, given scientific assumptions, an inexplicable central co-ordinator in the brain does seem very mysterious, even miraculous. If science can show us the workings of the brain why should we have to start all over again by having to explain in physical terms how they are controlled? *Ex hypothesi*, the workings of the brain are all there are.

Nevertheless the project of explaining consciousness hangs in the air. Who is explaining it? Who is convinced by the reasoning provided? Presumably self-conscious rational subjects have to weigh the scientific evidence and even perhaps be convinced by it. The arguments which demolish the self, and the idea of a rationality which cannot be reduced to physical processes, depend for their life on rational assessment. This has to be performed by those capable of a rational judgement that cannot simply be explained in scientific terms. That itself is just to push the problem back a stage, since it begs the question as to what is to count as an adequate explanation. The assessment also depends on a gathering and collating of information which suggests some 'central' co-ordination. Whatever physical processes are involved in the brain, they cannot be assumed merely to be competing. At some point their products must be compared, correlated and judged. Without something like what is derided as a Central Meaner, the idea of explaining consciousness could not be entertained. Without epistemological criteria for what constitutes good reasoning, major claims by science to explain our rational thinking could never be assessed.

All attempts to give science monopoly rights over human reason are bound to fail, since the reasons for doing so can never be adequately accounted for from within science. Those who are capable of being swayed by such arguments are capable of a rationality which underlies science but which cannot be wholly explained within it. Science cannot exist without the reasoning powers of scientists, and if they are seen to be something totally different, the whole status of science is put into question. The fear of 'homunculi' generating a regress of explanation leads many into a situation where they implicitly decry the possibility of rational judgement while being involved in precisely that. Because they assume that science can produce answers to any conceivable question, they put their confidence in science in a manner that could not be justified within science.

Dennett's approach is based on an unargued trust in science. He tells us:

> The concepts of computer science provide the crutches of imagination we need if we are to stumble across the *terra incognita* between our phenomenology as we know it by 'introspection' and our brains as science reveals them to us. (1991, p. 433)

He advocates that we think of our brains as information-processing systems, and considers the revelations of science as the touchstone. Yet why should science have the final word, and preference be given to what is in the public sphere? Some rational justification must be given for science to contradict what we each imagine we know from our own experience. Science, moreover, cannot exist in a world without selves or centres of consciousness who can make judgements about truth. Scientific progress depends on considerably more than the information-processing of a computer. Like other aspects of human endeavours it requires a reasoned and conscious weighing of alternatives, with even at times a creative leap in the dark beyond the evidence currently available. Science has not only not got a proper understanding of the conditions which make it possible. There is no way it could ever have.

The Question of Legitimacy

It would seem impossible for science to be able to explain its own occurrence, since no room is left for a justification of scientific forms of explanation in the first place. Yet again and again philosophers of a naturalistic inclination imagine that science is its own justification. Arthur Fine draws a parallel between the urge to give a justification for induction, which always appears to come up against the problem that only an inductive justification is forthcoming, and the problem of justifying the existence of the external world on the basis of science. He says: 'Only ordinary scientific inferences to existence will do, and yet none of this satisfies the demand for showing that the existent is really "out there"' (1986, p. 131). Yet in the case both of problems about realism and induction, there appears to be a search for something more than empirical discovery. Appeals to science seem to beg the question since the questions of discovering reality and generalizing from past to future seem to precede science, rather than simply arising within it. Science cannot be expected to solve questions which are not only beyond its capabilities, but for which it needs an answer in order to get started.

According to Fine 'the realist is chasing a phantom, and we cannot actually do more with regard to existence claims than follow scientific practice' (1986, p. 132). Justification could only come from within science. Anything else will be damned as metaphysics. Bas van Fraassen, writing as an empiricist, comments that 'realist yearnings were born among the mistaken ideals of traditional metaphysics' (1980, p. 23). Yet if we stay within science, we seem to be caught by a dilemma. When science itself is a product of human consciousness, scientific explanations of consciousness themselves presuppose what they are trying to explain. There seems at some level no escape from the assumption that human reasoning can itself grasp what is true. That does not stand in need of scientific explanation, and if it is so explained, the explanation itself claims the very truth that is at issue. Science itself depends for its potency on the assumption that it is attempting to reflect things as they are. We cannot accept it on its own terms, unless we understand what its purpose is. This, though, is immediately to involve us in wider questions about realism and induction. Science, like all reason, must be grounded in the way things are, and metaphysical reflection gives an indication of the kind of world which confronts the scientist. Otherwise we have no framework in which the scientific project can make sense and be legitimized.

It will still be urged that all reasoning has to come to an end somewhere. We do not wish to be involved in the paralysing kind of regress which so often appears, when an apparent explanation, like the inductive justification of induction, is seen to be no explanation at all. Can metaphysics itself avoid this? If it, in its turn, generates regresses, many will say that they might as well just take everyday scientific practices at face value. We can simply stop where we choose to, when there is no real reason for doing otherwise. Yet at the same time, it can be seen that if, say, the world *is* structured and behaves in an orderly way, this can ground our scientific practices as nothing else would. If we really do see things as they are, and this is not an accident, our reason is grounded properly in reality.

When human reasoning is thought to be simply a public and collective practice, matters are not made easier. Any collective social agreement, whether in science or in the political sphere, generates questions about its basis. An example is the question of the legitimacy of a state, an issue to which questions about the legitimacy of science are bound to be compared. How is it possible, for instance, to found a constitution legitimately? It is difficult to see how there could be any beginning unless something is taken as given. What counts as legitimate in a state depends on its constitution, and yet if it is to inspire respect, the drawing up of the constitution should not be a merely arbitrary affair. A constitutional convention needs a basis, and its membership has to be established in some definite way. The representative nature of the convention can only be established by a constitution, and yet that is what the convention has to draw up. Some political philosophers would have us draw a veil over origins. What matters now, it will be said, is the tradition that makes us what we are, and not where that tradition might be grounded. There are perhaps affinities between this and the pragmatist position in science. This may seem plausible in a society, like England, with a long history, but it seems irrelevant when, for example, there is a desire in a country to make a radical break with the past. There may, for instance, be a wish to move from a racialist to a non-racialist society. The problem in such a case is that either a constitutional convention would reflect in its composition the injustices of the previous racist society, or it already embodies non-racialist principles. Yet it was set up precisely to chart a course to the latter position. It is difficult to ground the writing of the constitution in anything, if the legitimacy of the state cannot depend on anything other than the fact of its own existence. There seems, as in science, to be a choice between grounding the constitution in something that immediately demands the same kind of justification as the constitution, or just accepting the constitution at face value.

One way out of the impasse in political philosophy is to make an appeal to something beyond the state, such as human rights. They could perhaps be seen as part of an objective reality which stands in no further need of justification but which does justify the way a democratic state is ordered. It certainly provides a firmer grounding than the mere fact of human agreement embodied in tradition and social convention. Agreement on its own will always seem somewhat arbitrary, as tradition and convention can produce many different types of society. Different groups can and do come to different conclusions. The invocation of human rights is an attempt to ground actual or potential constitutions in something more fundamental. The issue then moves on to whether such putative rights are real or themselves socially constructed in some way. Forms of argument appear parallel to those about the foundations of science. Indeed the two areas can converge, when science is seen, as by Rorty, as analogous to a political institution. Yet it is already a big mistake to view science in political terms as a human institution and nothing more. Indeed it is singularly unhelpful since, as we have seen, the question of the legitimacy of political institutions will not go away. Seeing science in political terms will not in the end avoid the question of legitimacy, particularly as legitimacy is itself a political concept. Unless politics is just a matter of the wielding of power, the issue of legitimacy cannot be shirked.

Agreement is important but it is always going to be very flimsy, unless it is based on the common acceptance of shared principles and shared beliefs about what is the case. It will otherwise be merely a matter of compromise, and will only last as long as it is politically expedient. Yet if this is so, the principles themselves cannot be seen as merely the product of agreement. That is why even somewhat cloudy and rhetorical appeals to human rights can carry considerable force, since they claim to offer a way of grounding constitutions in objective facts about humanity. The point is, however, that viewing science as a political institution of some kind is not an effective way of dodging questions of legitimacy and of grounding. It is in the political arena that these questions become even more acute.

Science and Self-Reference

Science cannot explain its own existence, let alone justify it. It is a product of conscious reason, and that cannot be reduced in scientific terms to a description of what happens in the brain. Such a scientific exercise can only succeed in undercutting itself. This is a situation that often recurs. An explanation for something is given, which, if true, would make it impossible to see how the explanation could be given in the first place. Scientists can sometimes implicitly claim truth for what they are saying, whilst making it seemingly impossible to talk of such epistemological categories as reasons and evidence, let alone truth. Whilst we may reasonably seek for an evolutionary explanation for the existence of consciousness, a scientific explanation for our individual conscious states will only raise the question of the status of our conscious acceptance of that explanation. This generates a regress of explanation, each explanation requiring conscious agreement which itself may generate another explanation and so on. Any physicalist programme which aims at explaining all aspects of consciousness in scientific terms will run up against this problem. It has to claim truth, and yet our recognition of truth cannot be encompassed in physicalist language. All scientific explanations must come to an end somewhere with a claim to truth. Science must always be grounded in reality, and human scientific ability depends on the ability to recognize truth. However sophisticated and flexible computers are, they cannot share this human characteristic. They are confined by their programming to perform the tasks set them. Some would say that humans are also constrained, because they have been in effect programmed by evolution. Yet saying this is itself to rest on the perceived truth of a scientific theory. It is unlikely that evolution has programmed us to believe in evolution. Once again the recognition and assertion of truth can be seen to be ineliminable.

Beliefs about what is the case cannot simply be replaced by, say, reference to neuronal events in the brain. The latter are not in themselves *about* anything. This is the problem of mental representation, which becomes a problem once it is thought that a theory has to be produced in terms approved by the natural sciences. Even if some would not go so far as to say such representation does not exist, they would certainly wish to claim that it plays no real role in a theory of the explanation of human behaviour. Jerry Fodor sums up these concerns when he holds that unless we give a naturalistic account of mental representation, there will be no place for such an idea in science. He says that the deepest motive for this comes 'from a certain ontological intuition that there is no place for intentional categories in a physicalistic view of the world, that the intentional can't be naturalized' (1987, p. 97). Such a view crops up again and again. It is indeed part of the general suspicion about consciousness. The idea that mental properties not only exist but have an influence in the world is dismissed simply on the grounds that it is unscientific. The spectre of dualism haunts every such discussion. Fodor takes it as axiomatic that 'whatever has causal powers is *ipso facto* material' (p. x). Science cannot understand any other kind of causal interaction, and so if beliefs cannot be translated into physical terms, they must be discarded. Yet this would seem to make impossible the holding of any philosophic position, even about the nature of causality. Further, if we do have beliefs but they have no causal role, this itself makes the power of philosophical theories seem mysterious.

It is a faith in science which encourages physicalism and prompts attacks on such views as dualism. Dennett admits he cannot show that the latter is false or incoherent (1991, p. 37). He simply maintains that 'given the way dualism wallows in mystery, *accepting dualism is giving up*'. We will certainly not find physical explanations if we do not look for them, but what is at issue is whether global physicalistic theory, stemming from an unargued trust in science, could ever be fully coherent. Like all global theories it has to be applied to itself, and it furthermore relies on the right of science to claim truth. In a sense, it wants the benefits of metaphysics without the disadvantages. It wants to be comprehensive in scope, going far beyond the empirical evidence. It behaves as if it is properly grounded in reality. Yet, at the same time, it wants nothing to do with non-physicalistic categories, which it is liable to deride as 'mystery'. At one point Dennett talks of 'those who would defend the Mind against Science' (p. 40). Yet this dichotomy suggests that we should choose Science instead of Mind. The flaw is that science is itself the product of mind, and, as a result, the mind can never be eliminated by science. This is not to suggest that dualism is necessarily acceptable. It is to say that science cannot challenge the reality of, say, our conscious reasoning processes without undercutting its own claims to truth. Scientific claims to truth gain their power precisely because they are understood in non-physicalist terms, as saying something *about* the world. They express the beliefs of the scientists involved about what is true. Once they are conceived as *only* the end product of complicated neuronal processes, they are facts about individual brains and nothing more.

Just as a political constitution seems to stand in need of some external validation, science cannot be content with scientific explanations. It needs some assurance that they are proper explanations in terms of the way the world is. Science must rest on something beyond itself. Otherwise science will be trying to justify science in a manner which brings back all the paradoxes of self-reference. It cannot be emphasized too much that scientists are not explicitly concerned with what it is reasonable to believe. When they deal with belief, they look at what beliefs are held, not at what ought to be.

It may indeed be reasonable to believe what a scientist says, but that is not a scientific fact. It is rather a conclusion *about* science. The view that we *ought* to rely on science alone, and that an anti-scientific stance is objectionable, is fundamentally unscientific. It is that not in the sense that there is anything shameful about holding it. It is simply not a claim that can be made either from science or with the resources of science. Even the split between facts and values begs an important question. Why should we put our trust in the world of facts as exemplified by science? Such an attitude surely itself stands in need of justification, and if one cannot be given, it must presumably itself be regarded as simply another 'value'.

Whatever the merits of dualism, ruling it out simply on the grounds of its non-scientific nature is far too easy. It even begs the question as to whether we must always rely on terms that are already acceptable to science. If *present* science defines reality, we cannot expect our knowledge to advance, or envisage a major breakthrough of the kind that occurred when classical mechanics was replaced by quantum mechanics. It may not be dualists who are 'giving up' but those who are so committed to the present state of science that they cannot allow it to confront questions about consciousness in a fresh way. Yet once it is accepted that the subjectivity of experience does not make it any less real, it is dubious whether science should ignore it. Perhaps science has got too caught up with the public world. It is indeed a paradox that empiricism, the philosophy which was most encouraged by science, placed its whole emphasis on the fact of human subjective experience. Our knowledge of the world was thought to be built up from the sights, sounds, tastes and sensations which help to constitute private experience. Now the science which was supposedly built on that empirical knowledge takes it upon itself to destroy the very idea of private experience. The urge to bring everything into the public sphere, and to emphasize the priority of theory rather than experience, can still overreach itself. Having turned on its empirical base and even 'explained' it, science still requires our trust. It still claims authority. No longer does it merely systematize our private experiences and correlate it with those of others, to give science an intersubjective foundation. It challenges the way we each of us see the world.

The minute a justification is demanded for the overthrow of what we previously took for granted, none can be given. Once global claims are made in the name of science, there is no way of retreating to higher ground in order to claim truth *for* science. Reason, truth, validity, warrantability are categories which may underpin science but they are certainly not part of it. Science needs truth, but if it wishes to be self-contained, and to dismiss what lies beyond as 'non-scientific', it cannot have it. Yet some will continue to claim that the threat of regress is always a feature of reasoning. The demand for justification will never end. Like children who tiresomely always respond to an explanation by asking in turn why that is so, all justification may seem to carry with it the possibility that a claim to truth will be queried. Every grounding will have to be grounded in something else. All reasoning has to come to an end somewhere, and some may allege that it might just as well be with the methods of science as anything else. Any reason that can be produced can always be challenged.

Lewis Carroll, who was a serious logician in addition to his other talents, gave a graphic illustration of this in a tale where each step in logical inference was challenged (Carroll, 1895). He used the example of the sides of a triangle and envisaged those propositions as follows:

(A)   Things that are equal to the same are equal to each other.

(B)   The two sides of this triangle are things that are equal to the same.

(Z)   The two sides of the triangle are equal to each other.

The first two statements may entail the third, but there seems to be a tacit proviso (C) to the effect that if (A) and (B) are true, so must (Z) be. Yet, once we accept (C), it is agreed, we do so on the basis of a further proviso (D). This would hold that if (A), (B) and (C) are true, (Z) must be true. Yet, now, it seems, we have to have grounds for accepting (D). We can clearly go on forever like this, accepting nothing except on the basis of yet another proviso. Nothing is accepted at face value, and every logical step requires further justification.

Thus the tortoise who in this tale sets about tormenting Achilles, deliberately generates a regress which the narrator says was not resolved when he passed some months afterwards. Nothing could be accepted without challenge, and so no argument could ever start. Is it then arbitrary where reasoning stops? This is, of course, precisely the kind of concern which leads philosophers to take some human practices as given. It is also the concern which leads some to see the need for grounding reason in the mind of God. In each case, the urge is to ground reason in something that does not need such grounding.

Why, then, cannot science itself be a sufficient grounding? Claims to truth can certainly always be challenged, and further grounding requested. A refusal, however, to accept *anything* as true is the quintessential sceptical position. It is not so much a demand for rationality as the undermining of it. A systematic refusal to recognize truth is a challenge to the notion. Science, however, has put itself in the position where if it is judged within its own terms, it cannot assert truth at all. Every time a scientific claim is made, according to physicalism, an explanation couched in scientific claims can be given for the claim. The reason why this creates a vicious regress is that at each stage the claim to truth is discounted. Epistemological questions concerning justification are set aside. It is not a question of wanting an explanation for evident truth, of asking why something is as it is. It is a question of explaining how and why each claim to truth comes to be made. Justification is ignored. Indeed the issue can never be faced since what is always in question is why I or anyone else *think* something justified, not why it is. We can never get to grips with the problem of its rational acceptability, but only with the fact that it *is* accepted. Epistemological norms are discarded by translating them into something else. The distinction between facts and values can easily mean the end of philosophy, since everything non-scientific can appear non-rational. The irony is, however, that the distinction itself is philosophic. The restriction of reason to the inner workings of science itself depends on a non-scientific conception of reason.

10
Reason and Metaphysics

Our Ability to Reason

In previous chapters, we have seen the danger of generating infinite regresses in reasoning. We must, if we are to avoid total scepticism or despair, ground our reason at some point. We must be willing in the end to recognize truth. Any judgement at all depends of this ability. We should not accept something arbitrarily on the grounds that we have to stop somewhere. Yet nowhere is it more important to recognize this than when it comes to questioning our most valued practices. The unthinking acceptance of science, with its criteria of truth, the willingness to play the game, is far from a conscious understanding that science is founded on the very nature of the world it investigates. It seems that we have to stand outside science in order to judge its validity. Even the judgement that science can explain everything presupposes this ability. It involves reasoning about science and not just reasoning within it.

Throughout this book, it has been emphasized that rationality cannot be identified with physical, social or other kinds of process. It cannot even simply be identified with the criteria of a current body of belief and practice. What goes on in the brain, what forces are at work in society, even what epistemological standards are acceptable, themselves need to be recognized by a human reason that is not totally constrained by its context. It is never enough to say 'this is what we do'. There may, indeed, be the further question whether we ought to be doing it. Perhaps more significant, though, is the realization that recognizing what we do already involves the ability to reason about ourselves. It suggests that we can see what is true, as well as be conditioned to accept certain things as true. We may not be able to attain the detachment of a divinity, but still be able to achieve a certain distance between ourselves and the light of our judgements. We are all subject to the influence of our surroundings, and our physical make-up. To assert that judgements about truth can be entirely accounted for in terms of those influences not only undermines the power of reason. It uses the very reason that is being discarded in a manner that is self-destructive. We shall in this chapter explore these matters.

The pursuit of truth has led many to imagine that they could somehow stand outside everything and pronounce on the nature of reality. The idea of a God's-eye view can lead to the idea that we are totally detached from the world we describe. We seem to be in charge of all we survey. No doubt the ever-accelerating growth in scientific knowledge gives credence to the belief that we can control nature and manipulate it to suit our purposes. Yet scientists do not always maintain this position. When they begin to apply science to the nature of human understanding, they recognize that we too are a part of reality that is being explained. Whether they try to give a scientific account of the workings of society, of human psychology, or of neural events, they are accepting that science should be able to explain not just human biological processes, but the very essence of human rationality. We must remain a part of the world we are trying to explain. Yet the paradox is that showing this itself presupposes an ability to extract ourselves from our context to some extent. Scientists may disagree about whether social, psychological or neural processes are the most important, but the ability to show the relevance of any of them to human thought itself involves a certain detachment from the system in question.

Any scientific theory is put forward on the tacit assumption that its proponents can recognize what is true, or at least what seems to be so. If the reply comes, as it inevitably will, that most scientists have been mistaken most of the time, it must be accepted that even the power to recognize falsehood is important. Believing something on the grounds of its truth, or not believing it because of its falsity, both depend on the ability to see what is the case because of the way things are. That is different from being constrained by some lucky accident to adapt a belief that happens to be true for extraneous reasons. The point of scientific theories is that they claim truth, and those putting them forward must do so for reasons explicitly connected with alleged truth. Science has been one of the great achievements of human rationality. Those who wish to equate scientific procedure with rationality as such are not wholly wrong. Science is reason in action, and if the power of human rationality is challenged or decided, the status of science itself must be put at risk. Yet if rationality itself is in turn apparently undermined by science, the situation rapidly becomes absurd. Science may be properly grounded and correct in, say, its physical account of the workings of the brain. Then the brain's activities cannot be wholly accounted for in scientific terms. Notions such as those of justification and evidence, of being properly grounded, have crept back in. On the other hand, we may rule out such epistemological ideas as unscientific, and restrict ourselves to the description of neural events. Then the science in question, neuroscience, cannot claim truth or hope to be rationally acceptable. Even its descriptive powers rely on some notion of accuracy, and so it is unclear how any science can even get started without implicitly relying on an appeal to truth.

The attraction of science has always depended on its claims to knowledge. Yet that demands justification, and scientists who are not independent of the system they are trying to describe will always encounter problems in giving a complete account. A full scientific explanation of human rationality will both be part of what it is describing and have at the same time to assume that it is not. The attaining of truth, even of a partial kind, suggests an ability to attain a viewpoint that is external to the system being surveyed. Repudiating this kind of God's-eye view can, however, suggest that we are merely part of a particular system, operating according to its rules. The system may be a physical one or a conceptual scheme, but whichever it is, if this is our position, the problem is how we could ever be able to recognize the fact. We could not, *ex hypothesi*, stand back from the scheme and show how it worked, any more then we can step outside the physical system and pretend that we were not physical.

The problem is that an ability to talk about systems, at whatever level, entails an ability to stand outside. If I talk about a physical system of which I am part, I imply that somehow I have the ability to transcend its processes and view them in a detached way. I cannot admit that this ability in itself is purely physical, without once again adapting a detached position and implying that at least at the next level I can talk *about* physical events, without simply producing another.

Anyone who is merely part of a system and behaves accordingly will never recognize the fact. Seeing the system as a system already involves stepping outside it. The same argument clearly applies to the conceptual schemes of societies. Someone living in a closed community may appear to be very rational according to its norms and standards. No one in such a society would see it as one society amongst others. They would simply take their life for granted. Seeing a way of life as simply one amongst many possible ones already requires a degree of detachment. Anyone who even possesses the concept of a society or a community has already taken a step back from their way of life. Instead of its customs being taken for granted, they are seen in their context. Yet the ability to relate anything to its context presupposes that those doing so are not themselves bound by the same context.

In chapter 5, exception was taken to a definition of reality in terms of its being known even by God. Nevertheless, we also saw that the idea of a God's-eye view may provide a necessary contrast to the limitations of our own particular conceptual scheme. It certainly shows the need for detachment from our own context. The idea may be important simply because of its insistence on the ability to apprehend truth rather than voice the prejudices of one's time and place. It may suggest that we can see what counts as evidence for truth and what does not. In that case, it must be logically impossible to repudiate the notion, since the very repudiation itself involves a claim to truth. We cannot even claim that we are the prisoners of a system unless we can see that that is so. We cannot do that except by extracting ourselves, at least partially, from it. The prisoners in Plato's cave, who saw only the shadows of models of 'real' things, did not understand how limited was their understanding of reality. Once we have recognized the inadequacy of our system of thought, or seen that it has competitors, we have already stepped beyond it. That is human reasoning at work.

The paradox is that we are grounded in a particular reality, of which we are part, and yet we somehow have the ability to abstract ourselves from our situation and apprehend at least something of what is true about ourselves and about the world we live in. We are not then just following the rules of our community. We are able to see, for instance, the validity of Gödel's argument in a way that a computer never could, and in a manner that does not just reflect the presuppositions of our society. The constraints on us are real enough, since we are born into a particular society and inherit its way of thinking. We are also born with biological tendencies that have themselves been honed through natural selection acting on human genes. Yet an attempt to explain our beliefs solely in terms of constraints such as these will always refute itself. Even the attempt to do so demonstrates the power of human reason to break free of its shackles. Even the knowledge that it is shackled represents a significant advance.

Physical science can never, it seems, explain its own existence. When it tries to, it merely proves that scientists can provide good grounds for beliefs and can thus overcome to some extent the physical and social limitations that undoubtedly exist. The existence of science, as a purportedly successful enterprise, is in fact the best possible argument against its own ability to explain everything. Human reason can ultimately break free to sift good from bad evidence, to see what constitutes justification and what does not. If everything is reduced to a level of explanation appropriate to one of the special sciences, whether physics or something else, the question still remains as to how that science was ever in a position to give a genuine explanation.

The very fact of epistemology, of the ability to recognize the difference between good and bad reasons for belief, gives a rational grounding for science. At the same time, we should be faced with a warning that science should not overreach itself and try to undermine epistemology. If epistemology without science is vacuous, science without epistemology is undirected and even self-contradictory. It will tell us what we ought to believe about the world while denying that there are any grounds on which to base beliefs. Yet epistemology provides us with the sign-posts we need in order to help us arrive at truth. It both presupposes that we can recognize truth and also that there is a truth to be recognized. An epistemology that cannot itself be grounded in the way things are is going to be a poor guide. Epistemology itself needs some form of metaphysics. The only alternative is simply to take the procedures of science as normative. This though is circular. We will trust science because it gives us knowledge and we will regard its pronouncements as sources of knowledge because we trust it. We need an assurance of a link between science and the way things are, between reason and reality. Above all we need an assurance that there is a reality to be known. If reality were unstructured and disorderly, indeterminate and fundamentally chaotic, science would be impossible. It is through metaphysical reflection that we see the preconditions of science.

The Grounding of Science

Just because genuine science would be impossible without an ordered and regular world which humans can to some extent get to know, does not prove that science *is* properly grounded. Its much vaunted success may be illusory. Why should it not be merely the conceptual scheme of a particular group of humans at a particular time? It is after all, as we have just indicated, a characteristic of those who cannot stand back from a conceptual scheme, that they accept it at face value as the only possible one. We may be so wedded to the scientific world-view that we cannot distance ourselves from it enough to see its limitations. Any assumption that science sets the standard for epistemology might be merely the product of a particular historical epoch. Indeed the more committed we are to science, and unable to countenance any alternatives, the more possible it is that we are in the grip of powerful social forces. This possibility is enough to draw attention to the fact that science should never be accepted at face value. It must be assessed in the light of rational standards that pre-existed it and underly it. Metaphysics cannot be made somehow secondary to science. If science is grounded in metaphysics and needs it as its justification, then metaphysical reasoning has to stand on its own.

Metaphysics has often been associated with the search for necessary truth. It might seem that if it is to provide a firm foundation, it must not itself be left standing in need of the kind of justification it attempts to offer other human practices. Many philosophers are ready to live without foundations. 'Foundationalism' in epistemology is certainly under fire. It is generally agreed that the firm empirical foundations searched for by empiricism may not be forthcoming. The idea, however, that there is no target for our beliefs, no purpose for our scientific investigation, no genuine object on which faith, whether scientific or whatever, can be fixed, suggests that all our reasoning is going to be unconstrained. There will be no difference between good or bad reasoning, justified or unjustified belief, or pseudo-science and the genuine article. When every rational suggestion is as good as any other, we can no longer talk of rationality. The threat of nihilism is as real as the threat of a narrow scientistic vision which insists on a monopoly of truth, and thereby removes the possibility of talking about truth.

The possibility of metaphysics suggests an escape route from nihilism and a grounding for science. It gives scope for our reasoning without limiting it in a self-contradictory manner. Yet metaphysics may still seem like wishful thinking to many, who will feel that it is safer to remain within the proven methods of science. Yet the whole point is that this is not an option, particularly once the pursuit of science is questioned. Science has to depend on metaphysics if it is not itself to be discredited. We have to know that there is work for science to do. One can certainly do science without indulging in metaphysical reflection, but then one is in the position of someone in the grip of a conceptual scheme and unable to see that there are alternatives.

Without metaphysical reflection, there would be no suitable intellectual resources on which to draw to answer criticisms and challenges. The danger of such blind allegiance is that it can easily crumble in the face of attack. The situation is not unlike that of some member of a remote tribe, immersed in long-established traditions, who is suddenly confronted with the technology of Western civilization. Once, however, one is in the position of explicitly championing science, as in naturalism and physicalism, one has already taken the step back from science which is necessary. Claims that *only* scientific procedures provide a proper basis for knowledge can never be made from within science. Yet some philosophers are strikingly reluctant to admit that their advocacy of a scientistic outlook is as much a defence of a metaphysical position as any they would wish to attack. For them metaphysics is something akin to sorcery or witchcraft, dabbling in occult powers beyond the scope of science.

As noted earlier, this attitude is exemplified in Dennett's desire to explain human consciousness. He is adamant that he will not appear to settle an argument through any appeal to 'miracle' or 'mystery'. One of the ground rules for his project is a refusal to resort to any 'wonder tissue'. He says:

> I will try to explain every puzzling feature of human consciousness within the framework of contemporary physical science: at no point will I make an appeal to inexplicable or unknown forces, substances or organic powers. (1991, p. 40)

His criterion is 'the conservative limits of standard science', and only as a last resort, he says, would he be willing to call for 'a revolution in materialism'. It is perhaps surprising that he can even concede this as a possibility. He makes a close link between materialism and a commitment to the ordinary standards of contemporary science. Anything else will be likely to involve appeals to miracles. As a methodology, there is something to be said for pressing contemporary science as far as it will go. Too quick an acceptance that we cannot explain everything can be stultifying. As a metaphysics, a belief in what the structure of the world is actually like, it begs too many questions. We are being asked to take on trust not just scientific method but the actual theories of contemporary science. This does not take into account that science may quite properly develop in ways we cannot now envisage. Indeed, unless we are willing to accept the limitations of our current scientific view, there is no way that science *can* develop. Too great an investment is our current beliefs can itself limit the potentiality for scientific advance. A metaphysics which stresses the possible gap between the character of reality and our contemporary understanding can actually provide the stimulus needed for further research.

Metaphysics may separate reality from particular conceptions of it, even if they are correct. In so doing, it also encompasses the subject who possesses knowledge or belief. It is no coincidence that attacks on metaphysics from Nietzsche onwards have often concentrated on the abolition of any distinction between subject and object. The status of the subject is as much an issue for any metaphysical position as that of the object. Questions about the nature of objective reality may appear pressing, but the metaphysical question of the subject cannot itself be easily dismissed. Indeed the knowing subject, the self, or whatever it is to be called, is as much a part of reality as anything it knows. The dissolution of the subject can be disastrous, not least because it makes metaphysics itself impossible. The disappearance of the rational knower into an ensemble of neural networks is but one example of this process. When there is no one left to reason, rationality itself has to be jettisoned.

No trust in contemporary science is going to be sufficient, since it will never be enough to refer to what is accepted, and there is bound to be a slide into talking of what can and cannot be accepted. A refusal to allow reference to miracles and mysteries may appear laudable and tough-minded. Yet this is to imply that nothing *can* be mysterious to science, because, in the end, science *will* explain everything. We have to keep our balance on a tightrope between being credulous about hidden powers on the one hand and on the other merely insisting that contemporary science determines what can be known. Once the latter position is weakened to refer to science as it one day will be, in a manner reminiscent of Peirce, the further we depart from actual science, and the more metaphysical the doctrine becomes.

The Rational Subject

Scientific explanations of the nature of the rational subject may be an ambition of some contemporary science, but it is fraught with danger. The aim of science would seem to be to demonstrate how brain activity can fully account for anything that seems to be within the province of mind. The disappearance of the subject within myriad physical processes may seem a legitimate scientific aim. Yet it is an aim held by subjects making judgements. The very possibility of science depends on the possibility of rational judgement. Whatever the involvement of physical processes in this – and they clearly are intricately involved – the deliberate reduction of reason to neural events, or whatever, can never take account of the fact that the reduction itself is made on the basis of the reason that is itself being explained. Validity is being claimed for arguments while at the same time it is envisaged that such reasoning can be fully explained in other terms. Yet the explanation depends for its cogency on an irreducible appeal to questions of evidence and justification. Science cannot squeeze out of our thought the idea of a subject which is able to reason.

The penchant for scientists to be 'objective' suggests that somehow they believe not just that science can grasp truth but that the world can somehow be laid out as it is in itself. Those who criticize the idea of a God's-eye view are right to be uneasy about this, precisely because it totally leaves out of account the position of the very theorist who can make such claims. Yet the danger is that once this is recognized, the world recedes and all the emphasis is placed on the conceptual constraints present. In reaction to the bleak prospect of an objective world without the judgement of subjects, we can find ourselves in the relativist position of seeing reality as merely a posit of particular conceptual schemes. What is needed is both an indication that there is an objective world to be discovered, and an acceptance that we are part of that world, and yet able to distance ourselves from it. As a result, we are able on occasions to see what is true and discuss what is false, to make rational judgements and to see the validity of arguments. We are rational subjects, not wholly reducible to the processes of the physical world, but not wholly separate from them either.

This will, for many, raise the unacceptable spectre of dualism, and reinforce their distrust of metaphysics. They will shudder at the thought of immaterial selves as the source of rationality. Those who refuse to countenance anything beyond the scope of science will certainly find that too high a price to pay. Yet the point is not that we are populating the universe with superfluous entities, but that the pursuit of science has to rest on certain assumptions, whether explicit or implicit. It must certainly be concerned with an objective world. Otherwise it is participating in a very elaborate form of novel writing. Yet its practice must only be possible if those engaged in it are able to reason towards conclusions. This rationality will very often not be deductive in form but will need its own leap of creativity. It will have sometimes to proceed from the known to the unknown and from the present to the future. As justification, it will have to invoke the character of the world. That means that the subject must be detachable to some extent from its physical and social context to see what is true both about the world and about its own place in it. The very existence of science as a successful route to knowledge is testimony to the distinction between subject and object. So far from being an unscientific concept, that of the subject lies at the very heart of science. It is what makes science possible. Without subjects, there could be no scientists, since scientists *are* rational subjects.

We must, though, be wary of making too quick an identification between the idea of a rational subject, as the source of judgement of truth and falsity, and that of an immaterial self. Metaphysics can make room for dualism, even of a Cartesian sort. This may seem an affront to those who would like to rule dualism out *a priori*. It is important, however, to realize that denials of dualism can also be a legitimate outcome of metaphysics. The latter should be faithful to the character of reality. The exercise of rationality may have a more intimate link with the brain than dualism can recognize. Some philosophers are eager in this connection to talk of emergent properties, which depend on other physical properties for their existence but still can claim their own level of explanation. This is a very different approach from that of full-blooded physicalist views which have tried to eliminate all non-physical terms. That undermines any claim to validity that their arguments might make.

More sophisticated views might be able to claim greater consistency than simple reductionism. The point is however that they would themselves be archetypal metaphysical views. Metaphysics as a discipline cannot be defined out of existence, or dismissed as irrelevant or meaningless. Everything we think or do presupposes a view of the world and our place in it. If we are rational, that view should be explicitly examined and consciously held. We must be prepared to justify it and give reasons why we think it true. Just as the anthropic principle points to what has to be the case, given that we, as creatures of a particular physical composition, are part of the universe, so other arguments can point to what has to be the case, given that we are able to produce rational arguments in the first place. Such arguments draw attention to the conditions which make our reasoning possible. If we come up with an argument that proves that all reasoning is not what it seems, we have thereby attacked the basis on which we depended to produce our argument in the first place.

The human ability to reason must be the starting-point for all our thinking. Once we try to deny it, or explain it away, we merely use the very ability we attack. Such an exercise in self-contradiction is worse than being involved in an infinite regress. In the latter case, we merely give an explanation in terms of something which then itself needs a similar explanation. We have thereby explained nothing. The railway announcer who apologizes for the lateness of the train at Oxford and gives as the excuse that it was late at Birmingham, has failed to give any satisfactory excuse at all. He may have successfully transferred the blame to another part of the railway network, but has given no real reason for the lateness of the train. This is unsatisfactory and, as a mode of argument, it is unilluminating. Many have felt that appeals to the self as some 'ghost in the machine' is similarly shifting the quest for explanation back a stage in just this illegitimate way. Yet the dissolution of the self does not act as a block on a tiresome regress of explanations at other levels. Instead, it appears to provide an explanation, but at the great cost of removing the very conditions necessary for giving any explanation. It undermines itself. The subject disappears and with it the possibility of truth being recognized. *Ex hypothesi*, there is no one left to reason. There is only the occurrence of physical processes.

Can science explain everything? It will always be averse to placing limits *a priori* to what can be achieved through the application of scientific reasoning and experiment. It certainly cannot, except on pain of an infinite and vicious regress, explain how it can explain everything. The preconditions necessary for science cannot be subsumed within science as part of its subject matter. In that case, its warrant for giving explanation is precisely what is being explained. The contradiction is clear. So far from being the antithesis of the scientific outlook, a metaphysical approach to reality, however conceived, is the only way in which science can be given any foundation. The physical sciences have always provided some of the most impressive examples of the human ability to reason towards truth. Science, though, cannot supplant reason any more than it can destroy truth. It needs them both too much.

Contingency

The suggestion that there could be an infinity of worlds, or a cycle of universes, demonstrates by its very strangeness how, in contrast, Western conceptions in general, and science in particular, are imbued with the idea that we confront one reality which is in a process of evolutionary development. The idea of a progression through time seems to rule out the possibility of infinite kinds of occurrence, or never-ending cycles of eternal recurrence. The distinction between views of linear and cyclical change runs deep, but it is perhaps no coincidence that the former is connected with Judaeo-Christian ideas of purpose in Creation. From that source it has also been adapted to the needs of an atheist philosophy such as Marxism. It is not surprising that Western science also bears the mark of this inheritance. This does not in any way serve to validate assumptions that are often unconsciously made in science about progress and development. It does indicate

that science, so far from challenging religious views, may paradoxically itself sometimes be depending on them.

This would certainly be very different from the normal perception of the relation between science and religion. They are often
conceived as rivals, with religion retreating before the ever en-
larging scope of scientific knowledge. God has, it has appeared,
too often been used as in a Greek tragedy as a 'deus ex machina',
to be swinging in from the heavens by a crane at the end of a
complex plot to sort everything out. God has been invoked simply
because our scientific understanding had given out. Yet with the
onward march of science, the risks of such a strategy are all too
obvious. The theistic explanation can be discarded. Gradually
the gaps in our knowledge which were filled somehow by an
appeal to divine mystery become plugged by scientific means.
Science seems to progress remorsely in the pursuit of a complete
explanation of the world. The religious attitude comes to seem
irrelevant, and indeed, was alleged to be *merely* an attitude, not
rooted in reality. The latter was the province of science.

This understanding of religion, judged from a scientific point
of view, means that religion cannot even claim the dignity of
being shown to be mistaken. The implicit assumption, however,
is that as science determines truth, traditional religion must be
discarded. Its imagery may live on as pictures we still cherish and
stories we enjoy telling, but its rational claims to insights about
the nature of reality have been subverted. Religion can no longer
be about the world. This is such a widespread assumption that it
could even claim to be one of the dominant ones of our culture.
Yet it reaches its conclusion far too easily, not least because it
does not fully appreciate how far religious views made modern
science possible. That is not to say that religion is thereby vindi-
cated, but it does indicate that the relations between science and
religion are much more complicated than is sometimes realized.
They are not just simple rivals. Indeed, just as a matter of history,
it seems that Christian theology provided an important background for the rise of modern natural science in the seventeenth
century.

It has been argued that the empirical methods of modern
science depend on the view that nature is contingent, that it did

not have to be as it is, but was produced by the voluntary activity of the Creator. M. B. Foster writes:

> The contingent is knowable only by sensuous experience. If therefore the contingent is essential to nature, experience must be indispensable to the science of nature . . . because knowledge by reason cannot be adequate to a nature which is essentially something more than an embodiment of form. (1934, p. 404)

The idea of contingency is in fact essential for science. The latter could never have achieved its success if it had continued to believe that the deepest truths of the universe could be unlocked through reason alone. As long as metaphysics is considered to be merely preoccupied with necessary truth, it is hardly surprising that it will be thought to be irrelevant to science. We saw in Chapter 8 how an example of scientific method without the concept of a contingent order is given by the Greeks' approach to science. Aristotle, for example, often acknowledges the relevance of the evidence of the human senses (1939. 270b). This was particularly the case when they provided corroboration for views independently arrived at. Nevertheless, he was under no doubt that the physical world operated according to rational principles which could be understood *a priori*, without the need for empirical investigation. He was sure that circular motion had to be primary and says: 'That which is complete is prior in nature to the incomplete, and the circle is a complete figure, whereas no straight line can be so' (269a). This piece of reasoning determined his whole approach to astronomical questions.

This world, if contingent, must be seen to be not the only possible one. There is, however, a fine line between allowing the idea of contingency to degenerate into one of randomness, and, on the other hand, letting an idea of purpose or order solidify into one of necessary truth. The notion of contingency used by modern science avoids both of these temptations. Aristotle, on the other hand, veered towards the idea of rationally established necessary truth. Christian theology bequeathed to science the idea that this world need not have been as it is, but that it is not the product of blind chance. The freedom and the purpose of the Creator together ensured that Aristotle's philosophy, on the other hand, was preoccupied with the notion of purpose, of final causation. He often said that 'nature makes nothing by chance'. Through reasoning, he concluded that the heavens had to be changeless, eternal, and in some sense divine. There was, he thought, only one possible universe and one possible cosmology. There was then no room left for any idea of design. There may be an intrinsic purpose in things, with natural tendencies imbued 'by nature', but this is far from any idea of deliberate creation, as indeed Aristotle's idea of the eternity of the physical universe underlines. Reason can establish what has to be the case, and has no real need of empirical science. We do not need to observe or to experiment in order to know what is necessary.

Modern cosmology tends to veer between two opposite poles. It hungers after a 'theory of everything' and is tempted to demonstrate through mathematics why the world has to be as it is. Yet it is also tempted into various versions of a 'many worlds' view which suggest that the visible universe is a chance conglomeration of order in a sea of infinite possibilities. Barrow points out the implications of such infinity: 'In an infinite world, anything that *can* happen will happen – infinitely often in fact – if the infinite world is exhaustively random' (1988, p. 206). There would thus be random initial conditions, indeed an infinite number of them, producing an infinite number of universes, or subsets of the one Universe. The visible universe, on this understanding, is untypical of the whole. At the cosmic level the apparent order and comprehensibility of the physical world cannot be taken for granted.

Neither a belief in total necessity nor in utter chance leaves much room for ordinary science. Apart from the fact that neither gives any explanation for the sheer existence of anything, the one reduces science to mathematics, and the other makes scientific discoveries look highly illusory. They are merely summaries of localized order in a universe that is not like that at all. In fact even visions of oscillating universes and many worlds may, according to their own assumptions, be importing too much order and balance into an area of ultimate disorder. Science needs order, but not the order of the mathematical equation. It assumes, rather, an orderliness in a contingent world. Things do not have to be as they are. Otherwise we would not need to find out through empirical discovery the way they happen to be structured. Yet without a level of arrangement and structure science could not get a grip on anything.

The problem of induction illustrates the way in which science has to assume a certain level of regularity and order in nature, but cannot understand from its own resources why they are there. As we have seen, empiricism has never been able to deal with this question. With his doctrines about scientific revolutions and the incommensurability of scientific theories, no one has done more to challenge the place of rationality in science than T. S. Kuhn. Yet it is intriguing that even he feels an urge for a rational grounding for induction. He notes that the whole problem arises from the recognition that we have no rational alternative to simply learning from experience. As a result, he says we want 'an explanation of the viability of the whole language-game that involves "induction" and underpins the form of life we have' (1983, p. 569). Kuhn's response is that 'to that question I attempt no answer, but I would like one'. He shares he says, 'Hume's itch'.

Science is in fact powerless to explain the apparent regularities it discovers. They are the foundations it has to take for granted as the basis for all explanations. Even theories which try to re-establish randomness and chance can only do so by being built on a science which has already assumed regularity. Once that is put in doubt, all scientific explanation has to be put in doubt. It would become an exercise in trying to make sense of what is ultimately senseless.

Science needs the notion of a continuing order, a cosmos, but it also assumes that this is something it has to look for. It cannot be established by reason alone. The world is contingent, and does not have to be as it is. Only the assumption of a contingent order can give empirical science the scope it needs, whilst giving it an assurance that there is an underlying rationality waiting to be uncovered. Contingency demands an empirical method. We actually have to observe and experiment to see the way in which the physical is constituted. The concept of contingent order saves science from the necessity of mathematics on the one hand and the chaos of total arbitrariness on the other.

Why is there Anything rather than Nothing?

Whenever the assumptions lying behind science are examined, we are forced back to the question of why anything at all actually exists. Anyone averse to metaphysics will be likely to say that it is not a legitimate question. Some philosophers may want to stop it being asked for no better reason than that they cannot answer it. Yet it not only appears perfectly meaningful, but seems to encapsulate a feeling that affects many people at times. It expresses an 'ontological anxiety', an anxiety about the mere fact of existence. Anyone may of course suggest that such an attitude is irrational, an indication of momentary disorientation. Yet however much we are shown to be part of a world that has evolved, the question is likely to arise at some point. Even if human existence is linked somehow with the initial conditions of the universe, we will still want to know why the initial conditions were like that or, more fundamentally, why the universe as such exists. Many would hold, or perhaps fear, that such a course of questioning will lead inexorably to God as the source of everything. Others maintain that positing God as Creator is another instance of pushing the questioning back a stage, without really giving an answer. God's existence may be alleged to be in some sense metaphysically necessary. The retort could well come that the universe itself might just as well claim that necessity.

Van Fraassen sums up the situation, as seen by Hume and his empiricist followers:

> Granted that the regress in causation or explanation must have a terminus, there is no reason why that should not be the universe itself. There is no reason to regard God as a more fitting terminus than the world. For if the world becomes intelligible only through reference to God's will, how should we understand God's will? And if we cannot understand God's will, why not stop with the universe, which we could not understand in the first place? (1980, p. 212)

Much depends on whether positing God as Creator provides a firmer explanation for everything than just accepting the brute fact of the existence of the world. Is the world to be understood as needing further explanation in a way that God does not? Logical necessity on its own could not have produced a world. There need not have been a universe. The mere fact that it exists does not prove that it had to. Even the fact that we are here to ask the question proves nothing about the inevitability of the universe. This is no doubt partly what is at issue concerning the Strong Anthropic Principle. Once it is thought that our appearance is closely linked with the initial conditions of the universe, there may appear to be some inevitability about the development of life in general, and humans in particular. The connection between life and the basic structure of the universe might even seem to indicate some form of necessity. This, however, must be completely wrong. For one thing, there is no such thing as inevitable progress. At any stage in the evolution of the universe and then of life, something could have happened to thwart it. There was no necessity that humans should develop, even if the initial conditions were highly favourable. At best, they were necessary but not sufficient conditions. Anything could have happened to change the course of events. We could all have been wiped out.

Nothing can take away from the sheer contingency of the world, and the apparent order in it. Whatever the demands of mathematics, there still has to be a world which actually exists, so as to follow them. Even if some theory established that there could have been only one set of initial conditions, there is still the problem why the theory was instantiated. Even if alternative forms of order proved to have been impossible, and that itself is a very large claim, it is still true that *this* ordered world need not have existed. Existence can never be logically necessary.

The same argument can be applied to God. God, if He exists, exists necessarily. That, though, is to say something about the character of His existence. He would not be dependent on any other being. It is far from saying that it is logically necessary that God exists. The latter statement means that there is no possible world in which God does not exist. Yet many would hold that it is all too possible for our world to exist as a matter of brute fact. Those who are wary of stepping into the realms of metaphysics are not obviously contradicting themselves in accepting the fact of the world, and our place in it, and yet refusing to look for any further justification, explanation or grounding. Nevertheless it is hard to suggest that the universe can possess the kind of metaphysical necessity or self-subsistence which theists attribute to God. For instance if it is finite with a beginning and an end, it may seem to demand an explanation which an eternal, infinite universe would not need. The idea of metaphysical necessity carries with it more baggage than the fact of existence. Ordinary physical objects change and do not have to be as they are. They are contingent. We have seen that the universe does not seem significantly different in this respect. Certainly change has been a constant feature of its development, and its character bears no relationship to the changeless heavens envisaged by Aristotle. Indeed, the dynamic, evolving nature of the universe is very different from the kind of changeless, eternal reality which traditional metaphysics looked to as the ultimate explanation for all things.

The battle between those who wish for something lying beyond the transitory, physical world and those who revel in apparent constant change is the most enduring in metaphysics. Indeed it is essentially a battle about the possibility of metaphysics. From the time of Plato's objections to Heraclitus' idea of flux, to the present-day disputes between realists and post-modernists, similar questions have been raised. One need not follow Plato in constructing new metaphysical worlds to see what is at stake. There are those who see no need to explain change or to see order and continuity in the midst of flux. There are others who search for a reality beneath the appearances, and look for understanding when all seems unintelligible. It is not a coincidence that such basic urges were not only the source of metaphysics but also provided the rationale for empirical science. The attack on metaphysics will also make science impossible.

Like science, metaphysics is also concerned with the character of reality. It may hanker after necessity as a grounding, but if things just happened by chance and could have been otherwise it is foolhardy for any metaphysics to pretend otherwise. If all existence is contingent and ungrounded, that is itself a metaphysical conclusion of some consequence. In fact, in the end, even those who vehemently oppose the possibility of metaphysics are themselves reaching metaphysical conclusions about what there is. Their great danger is in being too content with superficial appearances and fleeting judgements. They may be so anxious to avoid being rooted in some transcendental reality that they may regard those as constituting reality. They end up by not even being firmly rooted in the ordinary physical world which is our home; that disappears in a cloud of becoming. Statements about its nature which go beyond momentary experience have to be ruled out. The perennial problem is that in avoiding questions that lead to the airy uplands of metaphysics, we seem to have made the pursuit of science itself impossible.

Metaphysical questions sometimes do not have obvious answers. There is no simple equation, for example, between the pursuit of metaphysics and theism. Atheism is also a product of metaphysical reasoning. We may conclude that the world does have the inherent structure and order required for the pursuit of science, but accept this as a brute fact, in no need of further explanation. We could not after all be living in a world which was too disorderly to support life. What we most assuredly should not do is continue the practice of science, without any reflection on the nature of the conditions that make the practice possible. Such reflection cannot be merely conducted within science without begging the questions at issue. Science itself needs a rational basis of some kind. Without it we will be at the mercy of those who wish to question not just its preeminence in our culture. They want to destroy its right to claim knowledge at all. What is at stake is the very possibility of reason and truth.

References

Aristotle 1939: *De Coelo* [On the Heavens]. London: Loeb Classical Library.

Ayer, A. J. 1946: *Language, Truth and Logic*. London: Victor Gollancz.

Barrow, J. 1988: *The World Within the World*. Oxford: Oxford University Press.

Barrow, J. 1991: *Theories of Everything*. Oxford: Oxford University Press.

Barrow, J. and Tipler, F. J. 1986: *The Anthropic Cosmological Principle*. Oxford: Oxford University Press.

Carroll, L. 1985: What the Tortoise said to Achilles. *Mind*, 4, pp. 278–80.

Changeux, J. P. and Dehaene, S. 1989: Neuronal models of cognitive functions. *Cognition*, 33, pp. 63–109.

Churchland, P. 1981: Is determination self-refuting? *Mind*, 90, pp. 90–101.

Churchland, P. 1987: Epistemology in the age of neuroscience. *Journal of Philosophy*, 84, pp. 544–53.

Collins, H. M. 1983: The sociology of scientific knowledge. In K. Knorr-Cetina and M. Mulkay (eds), *Science Observed*, London: Sage.

Cross, A. 1990: *The Rhetoric of Science*. Cambridge, Mass.: Harvard University Press.

Cupitt, D. 1991: *What is a Story?* London: S.C.M. Press.

Davies, P. 1992: *The Mind of God*. London: Simon and Schuster.

Dennett, D. 1991: *Consciousness Explained*. Boston: Little, Brown and Co.

Descartes, R. 1991: *Philosophical Works*, volume 1. Trans. E. Haldane and C. D. T. Rees, London: Constable and Co.

Dewey, J. 1938: *Logic: The Theory of Inquiry*. New York: Henry Holt and Co.

Dewey, J. 1960: *The Quest for Certainty*. New York: Capricorn Books.

Dewey, J. 1977: *The Essential Writings.* Ed. D. Sidonsky, New York: Harper and Row.

Dummett, M. 1991: *The Logical Basis of Metaphysics.* London: Duckworth.

Encyclopaedia of Philosophy. London: Macmillan, 1967.

Farki, C. and Guth, A. H. 1987: An obstacle to creating a universe in the laboratory. *Physics Letters,* B 183, pp. 149–55.

Fine, A. 1986: *The Shaky Game.* Chicago: University of Chicago Press.

Fodor, J. 1987: *Pyschosemantics.* Cambridge, Mass.: MIT Press.

Foster, M. B. 1934: The Christian doctrine of creation and the rise of modern science. *Mind,* 43, pp. 446–68.

Fuller, S. 1988: *Social Epistemology.* Bloomington: Indiana University Press.

Gadamer, H.-G. 1975: *Truth and Method.* London: Sheed and Ward.

Giere, R. N. 1988: *Explaining Science.* Chicago: University of Chicago Press.

Hawking, S. 1980: Is the end in sight of theoretical physics? Inaugural lecture.

Hawking, S. 1988: *A Brief History of Time.* London: Bantam Press.

Heidegger, M. 1962: *Being and Time.* Trans. J. Macquarrie and E. Robinson. Oxford: Blackwell.

Hooker, C. A. 1987: *A Realistic Theory of Science.* Albany: State University of New York Press.

Hume, D. 1888: *Treatise on Human Nature.* Ed. L. A. Selby-Bigge (first edn). Oxford: Clarendon Press.

James, W. 1907: *Pragmatism.* London: Longman, Green and Co.

Kant, I. 1953: *Prolegomena to any Future Metaphysics.* Trans. P. G. Lucas. Manchester: Manchester University Press.

Kant, I. 1970: *Metaphysical Foundations of Natural Science.* Trans. J. Ellington. Indianapolis: Bobbs–Merrill.

Kneale, W. 1967: Scientific revolution for ever? *British Journal of the Philosophy of Science,* 19, pp. 27–42.

Knorr-Cetina, K. 1983: The ethnographic study of scientific work: towards a constructivist interpretation of science. In Knorr-Cetina and M. Mulkay (eds), *Science Observed,* London: Sage.

Kuhn, T. S. 1983: Rationality and theory-choice. *Journal of Philosophy,* 80, pp. 563–70.

Leslie, J. 1989: *Universes.* London: Routledge.

Lucas, J. 1970: *The Freedom of Will.* Oxford: Clarendon Press, 1970.

Lynch, M. et al. 1983: Temporal order in laboratory work. In K. Knorr-Cetina and M. Mulkay (eds), *Science Observed,* London: Sage.

Lyotard, J.-F. 1985: *The Postmodern Condition: A Report on Knowledge.* Trans. G. Bennington and B. Massumi. Manchester: Manchester University Press.

Margolis, J. 1986: *Pragmatism Without Foundation.* Oxford: Blackwell.

Margolis, J. 1991: *The Truth about Relativism.* Oxford: Blackwell.

Nagel, E. and Newman, J. R. 1959: *Gödel's Proof.* London: Routledge and Kegan Paul.

Neurath, M. and Cohen, R. S. (eds) 1973: *Otto Neurath, Empiricism and Sociology.* Dordrecht: Reidel.

Neurath, O. 1983: *Philosophical Papers 1913–1946.* Eds R. S. Cohen and M. Neurath. Dordrecht: Reidel.

Oldroyd, D. R. 1990: The deconstruction of the social construction of knowledge. *Social Studies of Science,* 20, pp. 638–57.

Peirce, C. S. 1931–58: *Collected Papers.* Cambridge, Mass.: Harvard University Press.

Peirce, C. S. 1957: *Essays in Philosophy of Science.* New York: Bobbs–Merrill.

Penrose, R. 1990: *The Emperor's New Mind.* London: Vintage.

Phillips, D. Z. 1988: *Faith after Foundationalism.* London: Routledge.

Pickering, A. 1990: Knowledge, practice and mere construction. *Social Studies of Science,* 20, pp. 682–279.

Pinch, T. 1990: Deconstructing Roth and Barrett. *Social Studies of Science,* 20, pp. 658–63.

Popper, K. 1972: *Objective Knowledge.* Oxford: Clarendon Press.

Popper, K. 1982: *The Open Universe.* London: Hutchinson.

Putnam, H. 1983: Realism and reason. In *Philosophical Papers,* volume 3, Cambridge: Cambridge University Press.

Putnam, H. 1987: *The Many Faces of Realism.* La Salle, Illinois: Open Court.

Putnam, H. 1990: *Realism with a Human Face.* Cambridge, Mass.: Harvard University Press.

Putnam, H. 1992: *Renewing Philosophy.* Cambridge, Mass.: Harvard University Press.

Quine, W. V. 1969: *Ontological Relativity.* New York: Columbia University Press.

Quine, W. V. 1990a: Let me accentuate the positive. In A. Malachowski (ed.), *Reading Rorty,* Oxford: Blackwell.

Quine, W. V. 1990b: *The Pursuit of Truth.* Cambridge, Mass.: Harvard University Press.

Quine, W. V. 1992: Structure and nature. *Journal of Philosophy,* 89, pp. 5–9.

Rescher, N. 1977: *Methodological Pragmatism.* Oxford: Blackwell.

Rescher, N. 1991: Conceptual idealism revisited. *Review of Metaphysics,* 44, pp. 495–523.

Rescher, N. 1992: *A System of Pragmatic Idealism, Volume 1: Human Knowledge in Idealistic Perspective.* Princeton: Princeton University Press.

Rorty, R. 1989a: *Contingency, Irony and Solidarity.* Cambridge: Cambridge University Press.

Rorty, R. 1989b: *Philosophy and the Mirror of Nature.* Oxford: Blackwell.

Rorty, R. 1991a: *Objectivity, Relativism and Truth, Philosophical Papers, Volume 1.* Cambridge: Cambridge University Press.

Rorty, R. 1991b: *Essays on Heidegger and Others, Philosophical Papers, Volume 2.* Cambridge: Cambridge University Press.

Roth, P. and Barrett, R. 1990: Deconstructing quarks. *Social Studies of Science,* 20, pp. 579–632.

Ruelle, D. 1991: *Chance and Chaos.* Princeton: Princeton University Press.

Ruse, M. and Wilson, E. O. 1986: Moral philosophy as applied science. *Philosophy,* 61, pp. 173–92.

Schilpp, P. 1951: *The Philosophy of John Dewey.* La Salle, Illinois: Open Court.

Searle, J. 1992: *The Rediscovery of the Mind.* Cambridge, Mass.: MIT Press.

Stewart, I. 1989: *Does God Play Dice?* London: Penguin.

Trigg, R. 1970: *Pain and Emotion.* Oxford: Clarendon Press.

Trigg, R. 1973: *Reason and Commitment.* Cambridge: Cambridge University Press.

Trigg, R. 1982: *The Shaping of Man: Philosophical Aspects of Sociobiology.* Oxford: Blackwell.

Trigg, R. 1985: *Understanding Social Science.* Oxford: Blackwell.

Trigg, R. 1988: *Ideas of Human Nature.* Oxford: Blackwell.

Trigg, R. 1989: *Reality at Risk: a Defence of Realism in Philosophy and the Sciences* (second edn). London: Harvester Press/Simon and Schuster.

Trigg, R. 1991: Wittgenstein and social science. In A. P. Griffiths (ed.), *Wittgenstein Centenary Esssays,* Cambridge: Cambridge University Press.

Van Fraassen, B. 1980: *The Scientific Image.* Oxford: Oxford University Press.

Walzer, M. 1983: *Spheres of Justice.* Oxford: Robertson.

Weinberg, S. 1993: *Dreams of a Final Theory.* London: Hutchinson Radius.

Williams, B. 1978: *Descartes: The Project of Pure Inquiry.* London: Penguin.

Williams, B. 1985: *Ethics and the Limits of Philosophy.* London: Collins.

Wittgenstein, L. 1953: *Philosophical Investigations.* Oxford: Blackwell.

Wittgenstein, L. 1961: *Tractatus Logico-Philosophicus.* Trans. D. F. Pears and B. F. McGuinness. London: Routledge and Kegan Paul.

Wittgenstein, L. 1969: *On Certainty.* Oxford: Blackwell.

Wittgenstein, L. 1980: *Culture and Value.* Oxford: Blackwell.

Woolgar, S. 1988: *Science: The Very Idea.* London: Tavistock.

Index

DATE DUE